PARTICIPATION
IN INDUSTRY

Notes on Contributors

Ken Alexander   Professor of Economics
University of Strathclyde
Glasgow

Peter Anthony   Senior Lecturer in Management
University College
Cardiff

Campbell Balfour   Senior Lecturer in Industrial Relations
University College
Cardiff

Fred Boggis   Senior Research Fellow
University of Strathclyde
Glasgow

Ken Jones   Director, Social Policy Unit
British Steel Corporation
London

George Thomason   Professor of Industrial Relations
University College
Cardiff

PARTICIPATION IN INDUSTRY

EDITED BY
CAMPBELL BALFOUR

CROOM HELM LONDON

FIRST PUBLISHED 1973
© 1973 BY CAMPBELL BALFOUR

CROOM HELM LTD
2–10 ST JOHNS ROAD LONDON SW11

ISBN 0-85664-030-1 HARDBACK
ISBN 0-85664-058-1 PAPERBACK

PRINTED IN GREAT BRITAIN
BY W & J MACKAY LIMITED, CHATHAM

Contents

INTRODUCTION *Campbell Balfour*   I

WORKERS' CO-OPERATIVES *Fred Boggis*   21

THE COAL INDUSTRY *Peter Anthony*   56

EMPLOYEE DIRECTORS IN THE BRITISH STEEL
   CORPORATION *T. Ken Jones*   83

SHIPBUILDING *K. J. W. Alexander*   108

WORKERS' PARTICIPATION IN PRIVATE ENTERPRISE
   ORGANISATIONS *G. F. Thomason*   138

WORKERS' PARTICIPATION IN WESTERN EUROPE
   Campbell Balfour   181

BIBLIOGRAPHY   213

INDEX   215

Introduction

The phrase 'participation in industry' can mean many things to many people. Some groups on the left will not be satisfied with any compromise short of 'workers' control' where the workers are the dominant power in the running of a factory. Some groups on the right are a mirror-image of the previous group; they see proposals for joint consultation with workers as 'the thin end of the wedge' and fear a workers' takeover leading to workers' control. Such groups see the question as one of politics and social attitudes; the propertyless wishing to seize control of industry, and the propertied claiming the right to do as they will with their own.

Both sets of views are unrealistic in contemporary Britain; we do not advocate 'workers' control' in the classic sense, nor do we advocate the mild form of joint consultation which some industrialists might offer as a placebo. We seek to describe, analyse and prescribe meaningful forms of participation in industry through which workers are consulted about, and in turn can influence, the factors which shape their working lives. Men and women are now fully-fledged citizens, consulted, cajoled and flattered by politicians, national and local, who seek office and power. Yet many workers have no power to influence their working lives in the area or factory where they spend most of their time. Arguments about the forms which participation should take vary, and are often preceded by critical statements expressing doubts that workers want participation at all. This argument may develop the theme (a) that the older demands were based on simpler forms of production and we now live in an economy dominated by large and complex organisations in which workers' participation can only be small; (b) that the most effective method of influencing industry is for workers to join trade unions and work to make them more effective; (c) that the worker no longer lives in a 'work-centered' society, where long working hours and acutely felt grievances made him anxious for change, but in a 'leisure-centered' society, where he simply regards his job as something to be done, then walks out of the factory and forgets it (there is a graphic account of this attitude in *Saturday Night and Sunday Morning*, by Alan Sillitoe).

One line of argument comes from the management side and advocates participation as a means of directing worker frustration and energy into constructive channels. This was originally a psychological theory based on the researches of Elton Mayo and his associates, but in more recent years it is linked with the idea of 'job satisfaction' leading to a better-integrated work force, less industrial trouble, better industrial morale and therefore higher productivity or efficiency.

We advocate participation for its own sake. Men and women should have the right or opportunity to realise their full potential in work, as well as in their domestic, leisure or civic roles. This phrase is used by politicians East and West, and could have a splendid sound ending as an echo. We want to penetrate the phrase and look at the workings of industrial participation in our contemporary society.

Many trade unionists and their officials would answer the preceding paragraph by arguing that we should not mistake the shadow of participation for the reality of power. Power, they say, can only be achieved through strong union organisation and effective bargaining power. While we support and encourage 100 per cent unionism, we also say that the trade union role is too narrow at the point of confrontation. As two writers put it recently, in criticising union leadership as well as management, the unions have tended to 'limit their interests to earnings' while there is a wide range of working conditions and amenities which badly need improvement.[1] A contemporary example may underline the point: we were carrying out a survey of low pay in hospitals and visited a hospital laundry. Though the wages were low, the complaints of the women workers were about the heat under the glass roof in summer, rising to well over eighty degrees. They had complained to management and union for three years, asking for fans to be installed. After three fruitless and hot summers they went on a half-day strike and got their fans. The moral is that an effective works council or a good shop steward would have channelled their grievances

more effectively than an overworked and geographically remote union official. The danger of the moral is that workers
will not participate in decisions if they get better results from
shouting 'All out, meet in the car park.' An extension of the
range of collective bargaining needs to run in harness with
works councils.

Workers find wider horizons than their daily work if they
take part in union activities inside or outside the plant. Yet
opportunities for advancement through the union structure are
necessarily limited. This means that many able workers can
never express their feelings, thoughts or ideas about their work
in any meaningful form. Indeed, with some unions or regions,
the union may appear as an oligarchy demonstrating Michel's
Iron Law of Increasing Oligarchy whereby the leadership becomes virtually self-perpetuating and intolerant of criticism.
Goldstein's book on the T&GWU shows the union in the late
1940s, depicted as 'an oligarchy masquerading in the trappings
of democracy'.[2] Other large unions were criticised as bureaucratic in this period.

Trade unions now favour the 'open style' of leadership and
advocate greater power for decision-making being given to the
shop floor and stewards. Yet this may leave numbers of workers
dissatisfied as to the contribution which they feel they can make.

The creation of works councils or committees would do
something to meet this need. The right to be consulted about
matters which affect the working life of the employee, and the
chance to contribute to the changing or improving of such
policies, as well as originating new proposals, should be a
recognised part of a person's employment contract.

The demand for participation in public discussions has
flowered vigorously in the past fifteen years; the Campaign for
Nuclear Disarmament, the Consumer Research movement,
tenants' societies, demands for information about development,
and planning, motorways, airports, education. People are less
content today to passively accept the directives of anonymous
politicians, whether national or local. Universities, traditionally
conservative and hierarchical in their political structure, have
changed dramatically in their attitude to the presence of non-
professorial staff and even students on various bodies and com-

mittees. To some extent, the staff have been borne along by the pressure of the student demand for representation and participation at all levels of authority.

It may be argued that much of the above demand comes from middle-class groups, were it not that strong currents have emerged in working-class circles. One current is the movement towards decentralisation of trade union power. There has been the shift from national negotiations to local plant bargaining, described in the Donovan Report as 'the two systems of industrial relations'.[3] The emergence of strong shop steward groupings and workplace bargaining has been due to various factors; the inflation of the past thirty years increasing the bargaining power of unions; the change from the defensive national agreement setting a minimum wage rate, to the forceful plant bargaining which built a superstructure of payments on the minimum wage and backed this by threats or use of strike action. This required the emergence of active shop stewards, as the understaffed unions could not cope with the many negotiations. The spread of informal bargaining has undermined the national agreements and weakened the ability of unions to effectively control their shop stewards.

Linked to these economic and political changes have been the sociological effects of changing authority. The sanctions of unemployment and poverty have largely gone with the underpinnings of social security, the lessened authority of the manager and the foreman, with the latter complaining that he is now merely a 'post-box for complaints'. New forms of work control and supervision are being evolved, positively encouraged by some firms and reluctantly adopted by others.

There is another line of critical thought which shows that the goals of the worker and those of the industrialist are irreconcilable. There will be a continuing struggle over the distribution of the product as the manager seeks to exploit the work force and increase his profits. This line of thought is pre-Marxist, but was developed by Marx and appears in various forms in modern Marxism. While Marxists have varying views on participation in industry, they tend to denounce any form short of 'workers' control' as class collaboration and prefer confrontation through trade unionism.

Allied to the modern Marxist approach is the concept of 'alienation'. Modern industry tends to be large-scale and bureaucratic. Management becomes anonymous and impersonal, while the continuing division of labour in mechanised processes removes worker skill and builds this into the machine process. The worker becomes de-skilled and de-humanised. His workgroups are elongated along the production line and their interaction becomes less frequent. Increasing boredom leads to alienation and loss of interest in work, greater readiness to strike or take other forms of industrial action.

On the side of management there are some employers who have, for different reasons, built up systems of worker participation such as the John Lewis Partnership and the Scott-Bader Commonwealth. One variant of this is the 'human relations' school, influenced by the work of Elton Mayo and his followers. They believed that raising the morale of the work group and improving the channels of communication between workers and managers would in turn raise the productivity of workers. The American Autoworkers have been critical of this approach and called it 'moo-cow' sociology. In any event, research into work groups has shown that the link between morale and output is not a clear-cut one. Groups can have low morale and high output and vice-versa.

Modern organisation theory lays less stress on measures of physical output and would define high morale and job satisfaction as part of the organisation's 'output'.

PARTICIPATION IN INDUSTRY: CONTEMPORARY VIEWS

The roots of the modern cry for participation are deep in history. Different writers will trace it from different points, but an early rhyme in English expresses the basic issue, the master–man, or master–servant relationship:

> When Adam delved, and Eve span,
> Who then was the gentleman?

The growth of industrialism, the division of labour, the need for capital investment, the growth of an industrialist, dynastic elite, stratified the master–man relationship and channeled workers' energies into politics and unionism.

Theories and statements about 'industrial democracy', 'workers' control', 'worker representation' and 'joint consultation' or 'joint management' are numerous, as any classified reading list will show, but the enduring strand is the desire to change the employee from a 'hand' to a participant, to allow him to use his faculties as a worker in the way that modern democracy enables him to act as a citizen.

'Industrial participation' is a meeting ground for many disciplines; politicians of the Right and Left, as diverse as General de Gaulle and Marshal Tito, find in it echoes of their particular views; economists hope that it will help to raise productivity, enhance morale and reduce strikes; sociologists and psychologists have probed at different levels in search of the answers to 'alienation', 'job satisfaction', 'authority and decision-making'; lawyers are interested in the effects of workers' councils on the structure of authority, control and ultimate responsibility; finally, theologians of many creeds have welcomed workers' councils in industry as an end to man as a workthing and an aid to his development as a person.

Unfortunately, 'participation' is also a meeting point for cliches, and some of the powerful speeches of politicians leave only tinkling brass behind them. We search here for words and ideas which can still have substance. Historically, groups of workers have insisted on their ability to run a workshop as well as the employer. In small workshops, they saw no need for his existence. This argument is put graphically in *The Ragged Trouser Philanthropist*. Lack of capital hindered the founding of workshops by employees and, significantly, the first developments were in the infant co-operative movement with the sale of basic foodstuffs.

The growth of large-scale industry changed socialist and radical thought in the direction of trade unions, which are already affected by full employment. Instead of a defensive 'consumptionist' role, the union is expected to help raise output while exercising wage restraint. Habits, customs and job practices might be hampering an industry; what should be the role of the union towards the new demands made on behalf of the public by the nationalised industry?

While the absence of keen and able trade unionists was undoubtedly a major reason for the poor showing of workers' participation, some critics argued that the lack of important issues deterred the more able shop stewards, who preferred to be active on union matters and collective bargaining. Some place the blame with management for clinging to traditional authority, failure to consult adequately before the meetings, lack of a suitable agenda, while lack of technical and adminis- trative experience on the part of the men is also criticised, though it is agreed that suitable educational or training courses might take a considerable time.

PROBLEM AREAS

The perennial problem is the role of the trade union. Those who argue that the union is the 'permanent opposition' in industry, the sole representative of the rights and grievances of the workers, see continuing difficulties over the division of func- tions between the union and the joint employer-employee committees. If, as is argued, all employees on the consultative committees are union members, one area of possible conflict is avoided. Yet this presupposes that they are conversant with union policy and agree with it. If they are loyal and active members of their union this is usually the case. But another problem might arise if there are several unions who differ on policies, or departments within a plant with opposed interests.

Concern is sometimes expressed at some consultative com- mittees setting up 'absenteeism' or 'discipline' committees, to deal with chronic absenteeism which may be reducing the plant's output drastically and thereby cutting the earnings of more conscientious workers. This happens from time to time in the coal industry, where pits may be placed on the 'in jeopardy' list and threatened with closure. As such closures in areas like South Wales or the North East mean that communities are deprived of their only industry, the NUM has allowed miners to sit on Pit Committees which have taken disciplinary action against habitual absentees.

The difficulty for the union is that if a joint committee recom- mends the dismissal of a member, would the role of the union be compromised if it did not take up the case? Some argue that joint committees should leave problems of discipline to management so that the union is not compromised in any way by having some of its leading members agreeing to such action. This tends to be the attitude of the American trade unions who argue that the job of management is to manage, while the job of the union is to protect its members and get the best possible terms for them.

One form of participation in recent years which had an effect on trade union policy was productivity bargaining. This took the form of a joint approach, usually begun with an approach by management to the unions, though sometimes the reverse action took place, to the questions of excessive overtime, overmanning, job flexibility, pace of work, restrictive practices. Finally, the pay level and structure was usually changed to give substantial increases.

Some unions encouraged productivity bargaining in the mid- and late 1960s as a means to raise earnings above the incomes policy 'norm', but some also saw the joint action as a means of widening employee participation in the work process. As joint consultation had frequently been criticised as dealing mainly with health, welfare and safety, or less elegantly 'tea and toilet paper', committee discussions which involved shop stewards in important decisions concerning most aspects of their work were welcome. However, since 1970 and the rise of unemployment to the 4 per cent or one million mark, trade unions and shop stewards take a much sourer view of productivity bargaining as leading to redundancies which would swell the unemployed total.

Schemes for participation in the future must be introduced and fostered in a favourable economic climate.

Looking back over the post-war period, how do the chief advocates of 'industrial democracy' see the developments and prospects? G. D. H. Cole, whose writings spanned half a century, argued in 1957 that, though public ownership had made the workers' job more secure, his status had remained unchanged.[4] Cole proposes that workers be given partnership status in the firm, that greater attention be paid to redundancy problems, and that these measures meet and be raised to a question of rights. This did mean that jobs are guaranteed. Given full employment, workers should be 'adequately mobile from job to job'. They should be kept on the firm's books, and retrained for a new job with pay. Appeals against redundancy or dismissal should be heard by 'a jury of his peers . . . his fellow workers'.

While Cole's argument rests on the major proposal that workers should have partnership status, we can look at the advances made towards his above proposals by 1972. While redundancies remain a serious problem, redundancy pay is awarded on a graded scale according to length of service. Workers are not kept on the firm's books for retraining but the government is now committed to a much larger programme of retraining with some pay. The Industrial Tribunals have heard appeals on redundancy since the late 1960s, while the Industrial Relations Act 1971 gives workers the right to appeal against unfair dismissals before the Tribunals. Though the trade unions object to the Act as a whole, and seldom use the Tribunals, over 2,000 workers had appealed against unfair dismissal in the first six months of the introduction of this particular clause with some success. Finally, the electricity supply industry has introduced a 'status' agreement.

THE POLITICS OF PARTICIPATION

As we shall see later when looking at the situation in Western Europe and in some other countries, schemes of workers' participation can be introduced through legislation or through negotiation. Apart from the nationalised industries, there has been little attempt to bring in such schemes in peacetime. There are signs, however, that legislation could be introduced through Parliament.

The Labour Party has been more active in discussing workers' participation, and its introduction in the nationalised industries, than has the Conservative Party. Yet they admit, in 1968, that while the need for 'industrial democracy' is 'stronger and more urgent than ever in the past' 'progress in devising new forms of popular control has been scant'.[5]

The National Executive argues that mergers are more frequent, so that redundancies need planning, that workers have

rights in this and other matters, that companies should widen
the range of collective bargaining to include dismissals, discipline, new methods, manpower and nationalisation. Finally
that companies should provide far more information to enable
workers to discuss the wider range of issues.

The Working Party Report on 'Industrial Democracy' ex-
pands the NEC statement considerably. One controversial point
is their belief in 'a single channel of representation', composed
entirely of trade union members, and not two separate systems,
collective bargaining and joint consultation, which has been the
standard procedure in industry where consultation took place.
They argue, and quote Dr McCarthy, that if the shop stewards
negotiate with management on wages and conditions, any
other committee on which they sit 'which cannot reach decisions, albeit informal ones, they regard as essentially an inferior
or inadequate substitute for proper negotiating machinery'.[6]

Those who have recent experience of collective bargaining
will agree with much of McCarthy's thesis. The writer has been
in industries where committees were set up in the 1960s to cope
with 'productivity bargaining', 'welfare', 'safety'. After a year
or so the shop steward members used to demand that the three
committees become a single committee as they became im-
patient at the artificial separation of topics on the agenda. This
fulfils the need for participation to be meaningful and for representatives to feel they make decisions which can be carried
out.

At the same time, Norwegian trade unionists would dis-
agree about the 'single channel'. They preach, and practise, a
separation between the issues on which the unions bargain, and
those which the workers' councils decide.

The Labour Party Executive itself seemed to be uncertain as
to the forms which industrial democracy should take. The NEC
spokesman, replying to the debate, said that 'joint consultation
has been a flop' as it created an artificial division between
consultation and negotiation. Yet in advocating a single channel
of communication, he admitted that the Report was deficient:
'we have got to work out in detail the structure of management/
worker representation at shop, plant, and where necessary,
national level.'

TRADE UNION EDUCATION

Whatever the range of issues to be jointly discussed turns out to be, and some of them may involve a wide knowledge of company accounts, technical processes, marketing and manpower policies, there will be need for intensive training and education courses. National and regional trade union officers could advise on certain issues, but they are thinly spread over industries, firms and areas, and have to devote most of their time to major collective bargains and administration.

Those with experience in trade union education, who see dozens of shop stewards in the year, know the size of the problem. Most stewards can get a grounding in economics, statistics, accounts, negotiations and industrial law in 24- or 32-day courses, on a weekly basis. The TUC Education Department has expanded its work considerably in recent years, but they are handicapped by lack of resources. It costs the TUC £200–£300 per course of 24–32 days, and this estimate acknowledges that the major part of the course cost is borne by public funds and by employers giving men their wages during their absence.

The pressure of NIRC fines and various legal costs has pressed hard on TUC and union funds, and education may be cut back at the expense of legal costs. It is a barren choice leading to a bleak future for industrial relations. There is a strong case for state bodies to support trade union education which trains shop stewards, in its entirety, and the arguments are put forward in the CIR Report on Industrial Relations Training.[7]

FACILITIES FOR SHOP STEWARDS

Even with greater opportunities for education, employers have to make certain basic facilities and opportunities. The CIR Report on Shop Stewards discovered that many of these were still lacking.[8] One office at least, with a telephone, should be available. Storage for records and books, access to typing and other services, time to attend meetings, consult members, and go to training courses without loss of pay. Such facilities are needed before participation can be conducted on any basis approaching equality.

THE CRITICS

The obstacles to participation set out above are not great; company organisation and decision-taking can be changed, information can be readily available to committees, rules can be drawn up about secret or confidential industrial information, shop stewards can be educated to evaluate accounts, assess manpower planning and widen the scope of negotiations.

The critics do not deny such claims directly. They prefer to turn the question round and ask 'is participation necessary?' They say that it has been introduced as a means 'of reconciling ideological contradictions in a symbolic fashion' rather than 'solving practical on-the-job problems'.[9] Yet the critics agree to some extent with the Labour Party Report; separation of consultation and negotiation is meaningless. The US unions do not favour formal participation, yet their plant grievance procedures handle most of the issues sought by 'industrial democrats'.

THE FUTURE OF PARTICIPATION

Criticism has to be faced and countered, effective machinery has to be devised. While we may be accused of advocating participation in industry without the backing of opinion poll or government election mandate, we feel there is a pressure for participation taking shape among workers. The evidence can be drawn from various sources, and the demand does exist in Britain today. Even if this statement were to be disbelieved, we would invite the scoffers to look to Western Europe. Britain joined eight other countries in the larger Common Market at the end of 1972. All the other countries have some form of workers' councils, some introduced by legislation and some by negotiation. Whichever method is chosen by Britain, membership of the Market will place some pressure on us in the direction of workers' councils.

Finally, in the near future, there is the important issue of co-operation or confrontation. The industrial relations scene, and the economy, has been dominated by confrontation; the mines, the railways, and the docks. This has had a bad effect

on the economy, and a continuation of confrontation can only
be damaging to all. This is not to argue a community of interest
between employer and worker, where there may be a natural
dichotomy. But discussion and negotiation are more rewarding
and less damaging.

An extension of participation in industry could be a stepping
stone into the future which, in turn, could become a bridge.
It is clear at the present time, after the failure of the Industrial
Relations Act, from 1971 to 1973, to improve matters, that
industry in Britain badly needs such bridges.

PARTICIPATION IN INDUSTRY

The writers in this book have dealt with industries or areas in
which they have special knowledge or experience. Workers'
co-operatives give us an opportunity to look at the historical
springs of industrial democracy and some contemporary developments. Coal, as the first of the great nationalised industries,
shows us the hopes, failures and possibilities of a strife-torn
industry with difficult industrial relations. Steel, the latest of
the nationalised industries and one of 'the commanding heights
of the economy' has run into more difficult times with rationalisation and redundancy, but has developed the interesting
experiment in worker directors, which may turn out to be one
of the first among many as EEC recommendations and practices
spread from our Common Market partners into British industry.

Shipbuilding is the third of the great basic industries and
has been going through a period of readjustment to changing
world demand and international competition. The Fairfields
experiment is described by an ex-Director, while the account of
the Upper Clyde 'work-in' represents a different trend in
industrial relations, as does the analysis of the Yugoslav shipyard.

There have been a number of developments in private
industry which are analysed by Professor Thomason, who sets
out much of the theoretical and legal framework to the modern
idea of participation. Finally, the account of the present state of
workers' councils in Western Europe shows the extent to which
the idea of participation has spread and taken root in Western
Europe. The British stand alone in their coolness to voluntary

or statutory arrangements while the tide of worker demands for
participation and representation grows over Western Europe
and now is felt increasingly, even in Britain.

Worker Co-operatives

It should be noted that a number of co-operative societies grew
out of strikes or bad working conditions; they were not started
to enrich the employees. Fred Boggis quotes one man as saying
that the weekly money was better in a co-operative, but the
important fact was that they were their own masters: 'We aren't
liable to be bullied and sworn at . . . we can feel that we are, in
a way, free men.' This was their strength and their weakness.
They had to survive in a capitalist, competitive society. The
idea of producers' co-operatives flowered in the 1880s, fusing
together the ideas of Robert Owen, French and Christian
Socialism.

Radicals, past and present, including the Webbs, have written of producers' co-operatives as belonging to the age of the
'hand-loom weaver'. Large-scale production demands large
amounts of capital; even a light engineering factory employing
500 men needs a considerable investment in machinery and
financial backing.

Though there were complaints of lack of management expertise, producers' co-operatives were developing the type of
management, and the attitudes to output and work disciplines,
that are much needed in the 1970s. There are lessons still valuable today and we may hear more of the old suggestion: 'The
Board of Management of the Society is elected by the members,
and the Board appoints the Chief Executive.' The lessons of
workers' co-operatives are not only relevant for Britain and
Western Europe. Co-ops are being developed in the new
countries of the world, as the United Nations Industrial Development Organisation has reported. The importance of these
societies in countries, such as in Asia and Africa, which seek an
alternative form of production which is neither capitalist nor
communist, is obvious. In some countries the high labour content and low capital input of some industries make them well
suited to producers' co-operatives, while in others, Governments
are ready to help such developments as highly effective in rural

areas where traditions of communal living and work are strong.

The Coal Industry

Coal was one of the first major industries to be nationalised and
the National Coal Board had a statutory duty to consult with
employee organisations. Peter Anthony traces the progress of
joint consultation from 1948, the change from consultation to
communication, and the doubts over the duality of roles which
representatives felt as individual members of the committee and
as members of a trade union.

The analysis of committees from colliery to national level
showed that, at the lower levels, management dominated, and
this domination increased over the years. At national level, the
tendency was in the opposite direction. Another trend was for
expressed conflict to diminish, as almost all items discussed resulted in neutrality rather than conflict or agreement. The strik-
ing fact is that the absence of conflict in formal consultation
went side by side with great changes in the coal industry, from
the introduction of power loading to pit closures. Peter Anthony
suggests that the absence of committee conflict was due to the
the choice of subjects for discussion which would not raise conflict and quotes Dahrendorf, 'the attempt to obliterate lines of
conflict by ready ideologies of harmony and unity in effect serve
to increase rather than decrease the violence of conflict manifestations.'

As other writers have pointed out, collective bargaining by
shop stewards is regarded as yielding greater benefits to workers
than does participation on a committee without any effective
power. The NUM sometimes withdrew from consultation committees, but never from the collective bargaining process.

It is argued that the NCB should accept that there is a plural-
ity of views in the coal industry and that conflict rather than
harmony is a part of the management-union relations in coal.
Collective bargaining is seen as the most effective method for
handling conflict, and elements of consultation might be built
into the system. This means a role for the State in a tripartite
structure. A framework of planning and a national agreement
would allow management and unions to reach a measure of
joint agreement on pay bargaining and participation in mana-

gerial decision-making. This approach has been recently mooted by the mine-workers.

Worker Directors in the British Steel Corporation

Ken Jones describes the important experiment in British Steel from the inside; the nomination, appointment and training of the worker directors, and the way in which the important question of whether such men should continue to hold union office or not and other role changes was considered. In his analysis of the BSC scheme, he shows the problems of defining 'employee', 'involvement' and 'management' and the role which conflict plays as opposed to theories which stress co-operation and play down the union role. This is the contemporary dispute between the pluristic school which recognises divergence of interests and the unitary or 'team spirit' school.

He suggests that worker participation can be introduced from the top of an organisation, in spite of obvious disadvantages, as being quicker and less liable to be hampered by management opposition. However, a joint problem solving approach would develop a more suitable system of participation, if unions agreed to co-operate. The question of union representation and multi-unionism is discussed, although research into the BSC scheme showed that most employees thought that the worker directors represented all employees, not just certain unions.

From this he develops the important question of the role conflict in a man attempting to be a representative voice of his workmates (but not *the* representative) and also acting in a managerial role. Although some MPs find no difficulty in riding two horses simultaneously, this is often difficult in an industrial situation. The good and bad points of the scheme are discussed, the criticism that worker directors become divorced from shop-floor opinion, although in BSC they have worked part-time at their normal jobs. Worker directors are now allowed to choose whether to carry on their union work or not, and their choice has varied. From cautious beginnings, the scheme has resulted in greater participation in boardroom decisions for the particular men, and their contribution to decision-making has increased over time and helped the decisions made.

In spite of introductory difficulties due to industrial reorganisation and industrial relations tensions, the scheme has had some success. It was supported by the Chairman of BSC, his labour staff and numbers of union officials. Further developments now seem possible, and as the TUC has given cautious approval to the idea of worker directors, the BSC scheme may serve as a prototype for other industries in Britain.

Shipbuilding

The influence of work technology on the prospects for participation in shipbuilding is brought out by Professor Alexander. The uneven nature of the demand for ships, boom periods interspersed with redundancies, have led to defensive or work practices being developed by the union. Different occupational groups have built up job practices to protect their members and these groups have rarely combined together into one cohesive group.

The Fairfields experiment in the mid-1960s was intended to test the possibilities of developing a better industrial relations atmosphere in shipbuilding on the Clyde. As Ken Alexander points out 'the extension of industrial democracy within Fairfields was primarily a response to the shipyard workers' desire to participate or control'. Nevertheless, there was an extension of participation by workers through 'a network of briefing groups', and the attendance of shop stewards on committees from the Executive Management Committee to job evaluation. There were two trade union leaders on the Board. The Fairfields experiment had some success in showing that participation yields results even in a difficult industry.

The Upper Clyde Work-in

The saga of the Upper Clyde is much better known than that of Fairfields and the success of the workers in saving their jobs led to a change in Government policy of support to industry, and a number of 'work-ins' imitating the Upper Clyde workers' tactics. Yet 'the "work-in" was a myth, but a very powerful myth.' It served its purpose and kept the yards open and completed the ships already begun. It was a powerful force in shaping public opinion, receiving support from all political parties in West Scotland, and it smoothed out craft differences and tensions in the yards, welding the labour force into a homogeneous and united group.

Ken Alexander finally looks at industrial democracy in a Yugoslav shipyard, and traces the system of committees which provide 'an excellent structure for democratic management'. Some aspects of the Yugoslav system, with its broad basis of democratic participation, might be developed in the UK, but this would need an extension of public ownership.

Participation and the Private Sector

Though the majority of writers in this book are dealing with the nationalised, co-operative or state-financed sectors, the greater part of firms in the British economy are in the private sector. Professor Thomason reminds us that 'shared decision' between employers and employees is the more likely development here than schemes of 'workers' control'. He discusses the various categories of participation. The first is joint problem-solving, where workers help management in the search for solutions, and involvement may lead to commitment; the second is joint consultation, which has risen and ebbed in popularity over the years, and is defined as the 'joint discussion of problems in which both or all parties have some common interest.' Issues may be trivial or important, and the process may become a management tool for passing on information and easing the way for difficult decisions.

The Glacier model of participation is discussed, where the employees participate in setting the rules which the law allows. While decisions of the Works Council must be unanimous, management can still effectively block any changes of which they disapprove. The Glacier system encourages the flow of communication, which sometimes places middle management in difficulties. The question of role is crucial, in that a man may be playing an executive role in one task and a representative role in another. This, however, is a problem found in other schemes for participation.

Single Channel Participation

George Thomason looks at the Factory and Productivity Council of GKN-Shotton, which combines trade union bargaining with joint problem-solving. The main council consists of eight representatives from both labour and management, which deals with a variety of subjects, some of which would normally come under the trade union for collective bargaining. Discipline sub-committees can dismiss workers. The Charter agreement has produced good results in output and better industrial relations.

Profit sharing and co-partnership schemes are discussed. Such schemes have not been widely adopted in Britain, although the John Lewis partnership is the best known.

There are legal difficulties in the way of greater worker participation in the shape of the Company's Articles of Association. If Company law is not changed, then all employees may be made shareholders in the way shown by the Scott-Bader Commonwealth.

The Role of the Union

This is crucial to most systems of participation, as the union confronts management over the more important issues, although on a narrow range. Collective bargaining might well develop on the lines pursued by US unions and negotiations may embrace a greater number of issues and areas. It could be argued that British management has far greater powers over dismissals than have the capitalistic US managers. The reaction to this has been fierce; sit-ins and work-ins have become part of the industrial folk-lore. Plants may be seized and, as one shop steward said, 'operate if we can, and occupy if we can't.' Thomason argues that the technique is not revolutionary, 'but merely an extension beyond the traditional level of penetration to challenge management decisions.'

Participation in Western Europe

Britain will find that all other eight countries in the EEC have some form of participation, from the weak works councils of France and Italy to the more highly developed systems of Germany and the Netherlands, with two-tier representation on the Supervisory and Management Boards. This latter form of representation appears to be the most likely to be adopted by member countries of the EEC and the question is being debated in Brussels. The British Government, as well as management and unions, will have to join in the debate.

Different countries in Western Europe have been developing their own systems of participation, some of which throw up new methods and ideas. The Norwegians have had some success with the establishment of a Co-operation Council, which organises meetings, training and research between labour and management. New forms of 'shop floor' participation are also being tried. The Danes have recently (1972) developed an interest in 'economic democracy' and worker representation on management boards, as well as proposals for the gradual transfer of assets to employees through a payroll tax. The Swedes have proposals for government directors to represent the public interest on boards of large companies.

The scene in Western Europe is one of increasing interest in the extension of workers' participation. Britain cannot stand apart from the rising swell of workers' expectations.

CAMPBELL BALFOUR

REFERENCES

1. W. W. Daniels and N. McIntosh, *The Right to Manage?*, Macdonald, London, 1972.
2. J. Goldstein, *The Government of British Trade Unions*, Allen & Unwin, London, 1949.
3. H.M.S.O., *Royal Commission on Trades Unions and Employers Associations* (Cmnd. 3623), London, 1968 (The Donovan Report).
4. G. D. H. Cole, *The Case for Industrial Partnership*, Macmillan, London, 1957.
5. Labour Party, 1968 Conference Report.
6. W. E. J. McCarthy, 'The Role of Shop Stewards', Research Paper No. 1, for the Royal Commission on Trades Unions and Employers Associations, para. 58.
7. Commission for Industrial Relations, *Industrial Relations Training*, Report No. 33, 1972.
8. Commission for Industrial Relations, *Facilities for Shop Stewards*, Report No. 17, 1971.
9. G. Strauss and E. Rosenstein, 'Workers Participation: A Critical View', *Industrial Relations*, Vol. 9, no. 2, February 1970, p. 199.

Workers' Co-operatives: A Vital Experiment

FRED BOGGIS

It was unemployment, exploitation and victimisation which drove workers in the latter part of the nineteenth century to seek an alternative to the ruthlessly competitive form of industrial life they detested. More than a hundred workers' co-operatives* were formed and virtually all foundered within a few years. A few co-operatives proved viable and survived into this century; even fewer continued in existence to the present day. The workers' co-operatives of the 1870s, '80s, and '90s were themselves the heirs to a tradition and an ideology which had been in formation since the 1820s and 1830s. From these experiments there is much to be learned about the preconditions for success and the causes of failure.

Workers' co-operatives are still being formed today. The persistence of the experiment suggests it should not be so readily dismissed as insignificant. While its economic impact has been slight there are important lessons for the building of industrial democracy in workers' co-operation. This neglected experiment in industrial democracy has relevance to our own 'half-socialist, half-decadent-capitalist' age, because some of the barriers which led to earlier failure are being eroded and the prospects for future experiments are improving. The reasons for forming present-day workers' co-operatives may not be so very different from those which were present during the nineteenth century. Unemployment persists, although its name may be changed to redundancy. The desire for job security still motivates men in work to stay in work. What has changed is the balance between the motivations of co-operators, the fund of experience on which they can draw and the level of economic and management sophistication amongst those who perpetuate the experiment.

---

* Enterprises which are owned and controlled by their employees who share the surplus.

ORIGINS IN CONFLICT

The workers' co-operatives in this country are mainly in the
Midlands, around Leicester, in the printing, shoemaking and
clothing trades. This 'historical survival' represents only a few
of the trades in which workers' co-operatives have been formed
all over the country. The diversity of the trades and the geographical dispersion of the experiments indicates a wider diffusion of the idea of workers' co-operation than its present
centralisation suggests. What is common to many of the experiments is the condition of industrial conflict from which they
emerged. In 1873, when the workers in the lock trade in Walsall
were on strike, their demand was for a 10 per cent increase and
the introduction of a uniform trade price list. Agreement was
apparently reached but after the return to work the employers
refused to implement the list prices. The dispute dragged on
with some workers returning to the firm in dispute until about
twenty strikers were left out. What happened to the hard core
of the strikers is recounted in this way:

> These (workers) were spotted and refused employ-
> ment; so the trades council, after paying strike pay for
> about 17 weeks, decided to recommend the union to sup-
> port the out-of-work men in forming a Co-operative Pad-
> lock Society. The initiative was left to the men, but they
> were helped largely by a gentleman in the town who used
> his influence on their behalf. Several unions including the
> locksmiths', took up shares or gave donations; and certain
> individuals in the town helped in the same way.

In Glasgow the Co-operative Cooperage Company, Limited
(1866) was the outcome of a strike. The Scottish Co-operative
Ironworks Society had its origin in the engineers short-time
movement of 1872. After the strike a number of the leaders were
discharged and this provided the impetus for setting up an
engineering and ship-repairing concern. The London Co-operative Leather Manufacturers' Society was similarly formed in
1890, to find employment for strikers boycotted by their employers, as was the Co-operative Cabinet Makers of Bradford
formed in the same year. A strike in August 1890 resulted in the

setting up of the Manchester Billiard Table Makers Ltd. in October 1890. The stick and cane makers of London's East End formed a society in December 1890 to employ men who had been out on strike for ten weeks 'against the wholesale introduction of boy labour into the trade'. The North Wales Quarries, Bethesda (1903) was formed as a result of a strike.

DECLINING AND SWEATED TRADES

In some trades unions were too weak for strike action to be contemplated by the workers. In these trades, the sweated trades, there are examples of co-operatives being formed. These were trades in secular decline in some cases. Nailmaking, for example, produced a number of societies amongst which two were in Dudley, the Industrial Nail Manufacturing society (1879) and the Midland Nail Makers' Association (1884). The Bromsgrove Nail Forgers' Society (1887) make this statement in their 1890 report:

> We have paid our work-people 10% more for wages than most in the trade, and 20% more than the sweater. In many cases we have defended the weak from the op- pression of the sweater; and your committee are confident, if the whole body of nailmakers would join your society, sweating would soon be a thing of the past.

The East End of London produced a variety of Co-operatives, as might be expected since it had well-documented sweated trades. The master bass dressers locked out their employees when they formed themselves into a trade union. The men took shares in the Co-operative Bass Dressers Ltd. (1889) which went into production. The advantage of the co-operative as seen by the workers was that '. . . it is not only the better weekly money that we get, but the fact that we are our own masters. We aren't liable to be bullied and sworn at and sweated like slaves. We can feel that we are, in a way, free men.' In furniture a Productive Co-operative Cabinetmakers Society (1889) and a Co-operative Bedroom Suite Manufacturing Society (1890) went into pro- duction in the East End after 'open-air meetings of men in the trade who knew full-well the terrible effects of sweating.'

A more isolated community, in Suffolk, turned towards a co-operative solution to its difficulties. A contemporary account records the facts:

> The mat manufacture . . . has centred itself . . . in the western portion of Suffolk, probably on account of the low price at which labour can be obtained. There is little doubt that most of the men, with the miserably low wages they were able at any time to earn were constantly in debt to the shopkeeper. In fact, the credit system is literally chronic in these parts; numbers of families are slaves to it all their lives through, and at the mercy of those who employ them.

The Long Melford Co-operative Mat Society began work hastily in 1887 because of 'rumours of the intention of their employers to discharge some of them'.

In the 1860s the shoemaking trade in Northamptonshire was still a domestic industry, but by the 1890s it had become a factory operation. As in Suffolk the abuse of truck trading was practised. Virtually the only employment in some villages was shoemaking. Boots and shoes were made on Government contract. The system worked in this way:

> The contractors employ persons in this (Wollaston) and neighbouring villages to give out the work to be done, and these persons, in a great many instances, keep shops for the sale of groceries, etc., which groceries the workmen have been under an obligation to purchase.[1]

The Northamptonshire Productive Society (1881) was intended to deal with the evasion of the Truck Act at Wollaston and a society to combat a similar problem at Raunds was registered in 1887. Truck was not the only problem for in the case of the Bozeat Industrial Boot and Shoe Manufacturing Society we learn that it 'was started with the object of preventing the further lowering of wages, by the workers employing themselves.'

UNEMPLOYMENT

Amongst the societies formed in the late nineteenth century

were those whose purpose was to find secure employment for
their members. A Co-operative Filesmiths' Society in Sheffield
(1886) 'was established by the Filesmiths' Union with the view
of finding employment for members who were unable to obtain
work elsewhere.' In London the Co-operative House Painters
and Decorators (1887) had as their purpose to find work for
unemployed tradesmen and 'to do good work on strictly co-
operative principles.' The Eccles Industrial Manufacturing
Society established in 1861 illustrates the ability of a workers'
co-operative to survive technological change and trade reces-
sions and still give continuity in employment. The society began
as a hand-loom weaving establishment but made the transition
to power-loom weaving. Similarly, it survived a cotton famine
and trade setbacks but it still provided employment for its
members.

IDEOLOGICAL DIVERSITY

From Thurso in Caithness where paving was quarried to
Holborn in London where theatres and music halls were gilded
the idea of workers' co-operation was accepted and acted upon.
Makers of portmanteaux and trunks, buckets and fenders,
knives and forks, watches and clocks, all found the co-operative
society an attractive solution to their problems.

The 1880s was a decade in which the ideas of workers' co-
operation were crystalised and a major propaganda effort took
these ideas to the workers. In 1882 the Co-operative Productive
Federation brought together the previously isolated workers'
societies and began an advocacy of co-operation in the North
and East Midlands which enlarged the movements' membership. A year later the Co-operative Aid Association began propaganda in the South of England. The case put by the propa-
gandists is epitomised by the thinking of E. O. Greening who
campaigned for the co-operative organisation of production in
order that:

(a) Workers might regain possession of the implements of
    production which they lost in the Industrial Revolution;
(b) The basic conception of democracy, namely, government

by the consent of the governed, should be established in
industry;

(c) The greatest common measure of liberty and freedom in
industry might be secured by this industrial self-deter-
mination;

(d) The status of the worker might be raised from wage-
earner to Conscious Co-partner;

(e) Pride of craft, largely destroyed by machine production,
might be restored in a workshop which engendered collective pride in product and organisation;

(f) Workers might participate in the surplus arising from
their associated endeavour;

(g) Consciousness of personal responsibility might be developed through worker direction and finance of the
undertaking.[2]

Greening's manifesto merges the thinking of French Socialism
with Owenism. To understand how this came about, it is necessary to briefly retrace the development of co-operative socialist
thought. Robert Owen and William Thompson advanced
theories about society and the economy which found ready acceptance amongst workers whose whole pattern of working and
family life had been altered by the change from domestic factory
manufacture at the end of the eighteenth century. The communities which Owen and Thompson argued for in the 1820s
were, as Smelser argues, regressive in character.[3] By reabsorbing
factory workers in self-supporting agricultural communities
with model factories and schools the reformers expected to produce ideal men. Labourers receiving their just reward in an
ideal environment would not fall victim to the vices of capitalism. Through the creation of simple communities which did not
require the complex division of labour of the capitalist system
it was expected that the profit-maker's role would be eliminated
and greater economic productivity achieved. 'Community upon
land' remained an evocative slogan into the 1850s, kept alive by
Owenite Missionaries and the growth of co-operative stores. The
distributive co-operative societies had their origins in community
enterprises protecting their members against high prices and
the adulteration of foodstuffs, but they were adapted to become

capital-creating institutions as well. From the funds accumulated through the sale of food to members it was expected to buy land and begin the community life. The more successful distributive societies, like the Rochdale Equitable Pioneers were able to accumulate sufficient funds to set their members to work in tailoring, bootmaking, corn milling, spinning and weaving. The larger productive units were separated from the Pioneers Society by the establishment in 1854 of the Rochdale Co-opera- tive Manufacturing Society. In this and other ways the Roch- dale co-operators accepted the differentiation of the market structure and the division of labour. They began to think not in terms of community-making but in terms of trade and manu- facture. The distinction between Producers' Co-operative Societies and Consumers' Co-operative Societies which has be- deviled the Co-operative Movement ever since was made mani- fest in 1854 by the Pioneers.

The other stream which mingled with Owenite Socialism had its source in France although it was an Englishman, J. M. F. Ludlow, who introduced the ideas into Britain. Ludlow provided the drive behind the Christian Socialist movement of the 1850s. While there was a distinguished theologian (F. M. Maurice) at the centre of the movement, which attracted the literary talents of Charles Kingsley and the legal skill of Edward Vansittart Neale, it was Ludlow's initiative which carried the spiritual fellowship along the path to Association.

In a recent study of the Christian Socialists Torben Christen- sen argues of Ludlow that 'he was not a systematic thinker inso- much as his views, to a large extent, sprang from and took form from the actual problems which, in his opinion, called for a solution.'[4] And the problems with which the Christian Socialists were confronted were firstly, those of pauperism, secondly, the condition of the sweated trades and lastly the conflict between labour and capital. The spiritual fellowship which grew up around Maurice in 1848 had no practical knowledge of the condition of the workers or what they were thinking; their attitudes towards labour problems were purely theoretical de- rived from the 'class' newspapers. Fortunately there had de- veloped a dialogue between the Maurice fellowship and some of the more literate and articulate workers. The views of the

Chartists and Owenites began to make an impression on a group of benevolent representatives of the working class who had been oblivious to the abundant and turbulent developments in the working-class movement since the 1832 Reform Act. Ludlow was in Paris which he knew since he was at University there, shortly after the 1848 Revolution and again in September 1849, when he visited *les associations ouvrières* established by the Parisian workers. From these experiences he was seized by the need:

(1) to Christianise Socialism by taking social problems seriously and to strive to remedy the evils under which the working class laboured. The main emphasis was on evangelisation among the workers to give them through Christianity what they expected from Socialism.

(2) to establish Associative Workshops as a means of combating pauperism and improving working conditions and wages.

(3) to establish home colonies on the land.

Ludlow's own study of the French Socialist thinkers was perfunctory. He did, however, learn something from Jules St. Andre le Chevalier who was expelled from France and joined Maurice's fellowship. Le Chevalier found Saint-Simon wanting as a social theorist but accepted Charles Fourier whose idea of '*l'association phalastenienne*' combined the interests of capitalist, worker and consumer. The *associations ouvrières* which Ludlow studied in Paris owed much to Philippe-Joseph-Benjamin Buchez, who was influenced by Saint-Simon but disagreed with his views on religion and the role of the state. Buchez placed his emphasis on voluntary action and association. In Buchez's view association would only succeed if based on Christian brotherhood—a doctrine which had appeal for Ludlow. Buchez took a practical approach to the creation of this new society by founding an association of cabinet makers in 1831 and this became a model for later workers' co-operatives.[5]

It required in fact a series of reports on sweating in the *Morning Chronicle* in 1849 written by Mayhew to shock the Christian Socialists into action.

Ludlow's response to this revelation of the extent of poverty and misery amongst the workers was to write an article in *Fraser's Magazine* in January 1850 which set out his views and then to agree with his friends to set up Working Associations for Journeymen Tailors. From the magazine article something of Maurice's opinions at this time can be established. He writes that charity, protection and emigration had been suggested as solutions to the problem of poverty but this did not expose the root of the problem which was competition which had eaten into the whole life and structure of society. The problem had to be tackled in a variety of ways. Parliament must enact legislation but this was not enough. The workers who had 'undersold each other as madly as the capitalists' had to take a part. The trade unions were not going to provide a solution. Rather it was association along the lines of the *associations ouvrières* in Paris which would obtain for the workers the full profits of their labour. Ludlow's appeal continues:

> Let this principle be applied not in one shape, but in a thousand . . . And let those who feel with me that the operative has need to be sustained in this effort to rescue him from his present thraldom, and that all who, wittingly or unwittingly, have contributed to bring him into that thraldom, be under the deepest responsibility towards him —let all such now contribute their Counsel, their funds, their custom, to further his deliverance.

This statement shows Ludlow moving further from his earlier reformist position. In 1848 he had seen virtue in a radical reform of the New Poor Law, because he saw the workhouse as fundamentally sound:

> It is our old English guarantee of labour. It is the pledge that work is to be found for every man who needs it, in great establishments, supported by local contributions where private employment fails.

While the existing institutions rightly attracted condemnation there was need for

> . . . an institution where agricultural and manufacturing

labour shall be carried on . . . for a sufficient, full remuneration, and under regular, intelligent and friendly discipline.
I wish it to become, in every parish, in the model-farm or
the model-factory, or rather both in one, as well as the
model-school, the model-congregation.

The echoes of Owenism are clear enough!

It was not community building but the formation of associations that the Christian Socialists embarked upon. They
spent a great deal of money and effort, in Raven's words, 'to
test by experiment the soundness and practicability of the
principle of co-operative production and to discover the best
methods of carrying it into effect.' The results of these experiments will be reviewed later. At this point suffice it to say that
it was on this theorising and practice that Greening was able to
draw for his own manifesto.

RECEPTIVE WORKERS

The propaganda of the Co-operative Productive Federation and
the Co-operative Aid Association found eager listeners, and in
the period between 1882 and 1903 the number of societies rose
from about twenty to more than a hundred and twenty. For the
boot, hosiery and clothing industry in Leicester and Northamptonshire it was a period of transition from domestic to factory
organisation. It was a period not without exploitation, strikes
and lockouts. While there were large factories opening in the
towns it was still possible for workers who preferred to work in
the villages and manage their own work to do so. When the
Kettering Co-operative Boot and Shoe Manufacturing Society
was started in 1888 the chairman reminded his audience that
'there was everything to encourage them, including the fact
that nearly all the present manufacturers in the town had themselves begun at the bench or the seat.' In terms of earnings,
especially during the starting-up phase of the societies, to work
in a workers' co-operative might mean lower earnings than in
the larger units. But there were other considerations—the style
of management needed in these workers' co-operatives was
clearly different. In the case of the Leicester Anchor Boot and

Shoe Productive Society (1892) there is an interesting aside in an anniversary history:

> Managing a Co-operative Productive Society is a difficult task. The freedom of expression which workers use towards those in authority makes it difficult for some natures to get along with them. The best side of workers needs to be appealed to, rather than exercising absolute authority or autocratic methods.[6]

The observations of a trade union official who visited the Leicester Co-operative Hosiery Manufacturing Society in about 1892 on the management style there are that:

> ... there is never any humbug here. The manager treats the work-hands as he would like to be treated himself. He is uniformly considerate and kind in all his dealings. The committee thoroughly approve this, and the consequence is that the work-people are quite a happy family. Another thing that makes the place such an enviable one is the fact that employment is very regular.

A member of the Co-operative Productive Federation at this time was the manager of a successful textile society, the Hebden Bridge Fustian Society, which had been formed in 1870 when that trade was turning from domestic to factory production. Although a trade union had been formed in 1860 low wages and irregular work were still taking their toll. Joseph Greenwood writes:

> The keen adversity through which we younger men had to struggle to provide the bare necessities for those dependent upon us was rendered all the more acute by uncertainty of our being able to stave off poverty and hunger, which in its turn brings on sickness and also premature death.

By starting a friendly and burial society which could also use its funds for trading purposes and setting up a cutting and dyeing establishment the workers were able, by starting in a small way, to build up, in time, a successful workers' co-operative.

As the earlier account of the diversity of societies in the 1880s and 1890s suggests, many were formed in trades where the capital that had to be committed was small and where it was possible to find suitable premises easily.* Experiences of this kind continued into the present century. Greening himself describes the circumstances in which the East London Toy Factory Limited was begun. 'When the war broke out in 1914, its first effect was to throw out of employment and reduce to semi-starvation large numbers of honest industrious working girls in the East-End businesses.' The suffragettes opened work-rooms for the unemployed, paying them 'a living wage instead of charity'. From this start grew the toy co-operative with advice from a Leicester workers' co-operative on rules, accounts and auditing.

A WRITE-OFF?

With only some two dozen societies still in existence today and a long history of failures in small-scale industry some observers have dismissed the workers' co-operatives as unimportant. Hugh Clegg says, 'Producers' co-operation can be written-off as a generally applicable device of industrial democracy; for in more than a century's experience it has at no time touched more than a negligible fraction of industry and it never will.'[7] P. Blumberg, who on most aspects of industrial democracy disagrees with Clegg, reaches similar conclusions:

> Producers' co-operatives, which do not involve workers significantly in management, have repeatedly been proved both economically and socially an inappropriate vehicle for workers' management. . . . In the Western world, they are economically inconsequential, especially when compared to the flourishing consumers' co-operative movement

Robertson and Dennison write more breezily that

* The Co-operative Typewriters' Society (1896) which provided a copying, duplicating, shorthand and translating service can have required little capital to begin its operations.

productive co-operation has not revolutionised industry, nor is it likely to do so. The main stream, not only of Capitalism, but of Co-operation and of Socialism, has swept past these gallant little craft; but in their limited sphere they have a vitality and an experimental value of their own.[9]

Verdicts such as these are not new. Harney, a leading Chartist writing in *The Red Republican* in 1850 of the proposal to start Associations says:

Even though but ten men out of ten thousand should be redeemed from the bondage of wage-slavery, it is so much good done, and so much evil dispelled. But do I believe that these associations can be made instrumental to redeem the great body of the people from their bondage to the capitalists? Nothing of the sort. Before I can be brought to that belief . . . I must be shown how working men are to procure the capital necessary to enable them to become proprietors of land, mines, railways, factories, foundries and establishments like those of the great builders of the metropolis.

Beatrice Webb, as the propagandist of Fabian Collectivism, was consistently scathing in her references to workers' co-operatives, and dismissed them as out of date:

The ideal of associations of producers belongs essentially to the time when industry was carried on mainly by hand labour in domestic establishments. We need not dispute the possible educational advantages of the self-governing workshop. Steam and machinery have killed it as certainly as they have exterminated the hand-loom weaver.[10]

After so whole-hearted a recommendation to consign workers' co-operatives to the dust-heap of history, what remains to be said? A more careful review of past failures, the present reality and future prospects suggests there is a more optimistic conclusion to be drawn.

REASONS FOR FAILURE

The Christian Socialists made their own analysis of the reasons for failure amongst the associations they had backed. There were associations formed amongst tailors, shoemakers, printers, builders, engineers, bakers, weavers, smiths, and needlewomen. The body which financed and overlooked the affairs of the associations was the Society for Promoting Working-men's Associations. The Promoters were the middle-class reformers, the professional men who had grouped themselves around Maurice and who were led by Ludlow. Two years after the Society's formation with many of the Associations defunct or in difficulties the Promoters announced in their Report that it was to discourage 'advances of money to working men about to start in association, unless they have first shown some signs of preparedness for the change from their own life, and have subscribed some funds of their own.' The Report then continues:

> Working men in general are not fit for association. They come into it with the idea that it is to fill their pockets and lighten their work at once, and that every man in an association is to be his own master. They find their mistake in the first month or two, and then set to quarrelling with everybody connected with the association, but more especially with their manager, and after much bad blood has been roused, the association breaks up insolvent, has to be reformed under stringent rules and after the expulsion of the refractory members.[11]

The first two points to be made about this statement concern recruitment. Firstly, the Promoters found it extremely difficult to get workers interested in their schemes, particularly skilled workers in the better-paid trades where trade union organisation afforded some protection. The uncertainty of Associative Production acted as a deterrent to those workers whose contribution might have been significant. Secondly, amongst those workers who showed interest no enquiry was made about their fitness as craftsmen or their attitude towards the principles of Association. It is a wonder that the Associations lasted as long as they did without intervention from the Promoters! The

Promoters were busiest in London but in Southampton and Manchester tailors and hatters who themselves formed Associations and did not suffer from the attention of the Promoters survived for longer, in the case of the Manchester Working Hatters Association lasting for twenty years.

The Report refers to quarrels amongst the Associates. A member of the Society who himself analysed the reasons for failures amongst the Associations also mentions that: 'The great evil is too many disputes, too many discussions, too many meetings, too much interference'. Given that the workers had only a faint notion of what Associationism was about and had little, if any, experience in the functioning of democracy is it surprising that they met frequently? The 'lessons' of the experiment seem obvious today.

Another observation from the same member of the Society was that: 'Dishonesty on one side and insubordination on the other were the real causes of failure.' This leads us into the area of management controls and industrial discipline and it is as well to look at later experience in this respect too.

Shop floor workers are production-orientated, yet the necessity, for example, for control and marketing functions is only grudgingly accepted. To take nineteenth-century workers and expect them to create a viable costing system, purchasing system and marketing arrangements for industries where competition was keen already was to expect too much. What happened to the London Portmanteau, Trunk and Bag Makers' Co-operative Productive Society will illustrate this point. The Society found it difficult during the first year to secure orders 'owing to the shopkeepers' repugnance to co-operation'. When the Society's books were audited they were found in such a muddle that it was impossible to give any idea of the position of the society. It was found later that manufacturing stocks had been pledged at pawnshops for sums below the cost of purchase. The manager left and set up in business for himself. The story is familiar to anyone who has examined small business bankruptcies!

During the difficult early stages of a business a contract is often an attractive proposition—it fills the order book quickly. The importance of correct estimation is imperative however if the business is to survive. One of the Associations went to the wall because it produced a quotation for a contract at too low a price. The Southwark Working Engineers had been formed after the 1852 Engineering lock-out by members of the Amalgamated Society of Engineers with money from the Christian Socialists. It apparently worked well enough as a production unit but its costings were wrong and it closed in 1854. The London Central Association of House Painters and Decorators obtained a contract to decorate a chapel in Wales but it was tendered for without being seen; the work was more than was anticipated and the association suffered a set-back.

The Leicester Hosiery Society to which reference has already been made was in fact the second to be formed with that name. Of its predecessor society the manager of its successful successor said:

> The committee of the Trades Union (which set up the Society), who of course were all framework knitters, had the management. They had plenty of knowledge how to make the goods, but they lacked commercial knowledge, what to make, and how to sell.

The Frameworkers and Gilders' Association did not react quickly enough to a change in taste for polished and painted wood for mirrors rather than gilding. Production for stock continued until the storage space was full and the goods could only be sold, if at all, at very low prices. The Association also lost customers at the top end of the market and found it difficult to produce cheap work in volume.

Contemporary observers of these societies were quick to point out, like Ben Jones, that:

> Working-men often overlook the absolute necessity for good book-keeping; not only the portion which is necessary to make out a balance sheet, but still more the portion which is necessary to enable the manager and the committee to know, week by week, whether profits are being made or not.

Similarly the advice was that:

> They must learn to produce not what they know, or think,

are the best goods, but those which the public requires, regardless of their intrinsic value.

But how was this knowledge of business management to be acquired? Possibly through prior experience in a retail co-operative society? Perhaps, but it seems an unlikely way to develop the skills required. Did working men have access to management education programmes in 1890? Hardly, since they have only been generally available in this country since the 1950s. The exasperation of the brilliant young Beatrice Potter when confronted with 'this dismal record of repeated failure: want of capital, want of custom and absence of administrative discipline', is, perhaps, understandable. Her writings show as little perception of the difficulties to be overcome by working men in starting and running a business as did the more idealistic Christian Socialists. Amongst the contemporary observers it is probably Alfred Marshall who shows most appreciation of the real nature of the problem: How does one find good management for these societies? Marshall appreciated that in an ideal workers' co-operative the employees would be:

> ... the employers and masters of their own managers and foremen; have fairly good means of judging whether the work of engineering the business is conducted honestly and efficiently, and have the best possible opportunities for detecting any laxity or incompetence in its detailed administration.

But the system would have difficulties:

> For human nature being what it is, the employees are not always the best possible masters of their own foremen and managers; jealousies and frettings at reproof are apt to act like sand, that has got mixed with the oil in the bearings of a great and complex machinery. And in particular, since the hardest work of business management is generally that which makes the least outward show, those who work with their hands are apt to underrate the intensity of the training involved in the highest work of engineering the business, and to grudge it being paid at anything like as high a rate as it could earn elsewhere. And in fact the managers of a

Co-operative Society seldom have the alertness, the inven-
tiveness and the ready versatility of the ablest of those men
who have been selected by the struggle for survival, and
have been trained by the perfectly free and unfettered
responsibility of private business.

Marshall saw the ambitious individual's path to success being
through the small business, the private firms or public company
but he concedes that:

> . . . co-operation has a special charm for those in whose
> tempers the social element is stronger, and who desire not
> to separate themselves from their old comrades but work
> among them as their leaders. Its aspirations may in some
> respects be higher than its practice; but it undoubtedly does
> rest in a great measure on ethical motives.[12]

The picture of the qualities required in a manager in a workers'
co-operative are beginning to emerge. Benjamin Jones lists
his qualities as being, 'a mixture of flexibility and toughness,
. . . and capacity for adapting himself to circumstances . . .
tolerant of other people's weaknesses and shortcomings.' In the
1890s the practical experience of the workers' co-operatives had
shown the need for a management style which only really
gained recognition in the social sciences in the 1950s.

INDIVIDUALIST!

In the world of Beatrice Potter the term individualist was one
of contempt. She described the workers' co-operatives as individualist associations of producers. Her solutions to society's
problems were always prefaced by the word collective. Collective bargaining was the solution to the wages problem, collective ownership by the nation or the municipality was the
way in which to supply the essential raw materials and services
the nation required while its food and clothing was to come
from the federally-owned factories of the retail co-operatives
owned in their turn by the nation's consumers. A tidy solution
which left no place for workers' co-operatives. In any case
argued Miss Potter:

Co-operative workshops are frequently established in bad times, or in decaying industries; they are formed to resist a reduction of wages, or to supply work to the unemployed. Obviously these associations are foredoomed to failure . . . Workshops for the unemployed may or may not be a wise Poor Law policy. But these workshops cannot become a living and independent part of a commercial system based on the interchange of commodities.

Apart from this criticism Miss Potter had other criticisms to make. She considered that the motivation of workers 'to increased effort and more sustained diligence' by the promise of a share in profits was not ethical. 'Working men are to be induced to work well and skilfully, not because they will help others by so doing but under this system of industry the workers will reap the advantages and suffer the consequences of efficient or inefficient labour.' This in Miss Potter's view was 'un-moral'. There seems some confusion here and it stems from both misconception and misunderstanding. Christian Socialism, which bears the brunt of her attack, certainly did not encourage profit-sharing. When it was suggested to Maurice that workers had joined the Associations because they 'wanted to make a profit' and that businessmen should be given control over their operations he recoiled in horror. His reply was: 'Talk as much as you like about putting the hand to the plough and drawing back. I never did put my hand to this plough!' A glance at Greening's manifesto shows what importance share in *surplus* has in the scheme; it is minor and it arises because there are always unanticipated gains and losses in any operation from innovation, changes in raw material prices, variations in loading and productivity, the responsiveness of markets etc. The basic clauses in the manifesto concern industrial democracy, personal responsibility and status ownership. But Miss Potter disregarded this view and labelled the Associations of Producers who competed amongst themselves as 'profit-seekers' intent on securing a large margin between the cost of production and the price given. The idea of consumer choice operating in such a way as to favour the products of one producer rather than another was too terrible to contemplate! Beatrice Potter was convinced that 'the new industry, with the subordination of the individual worker to masses of capital directed, by expert intelligence' was here to stay and 'the area in which increased diligence on the part of the workers can counteract the lack of capital, discipline and commercial knowledge is daily contracting.' To this view of the inevitable contraction of the sector in which workers' co-operatives can operate I will return later; suffice it to say here that I doubt its validity.

CURRENT APPROACHES

The disappointing experiences of workers' co-operatives in the nineteenth and twentieth centuries have been distilled into a body of know-how which can guide workers currently considering the formation of a society. The major respository of knowledge in this country is the Co-operative Productive Federation* (CPF) which was formed to protect and promote the interests of Co-operative Co-partnership enterprises in 1882. From its long experience of the problems confronting new societies the Federation offers advice and services which can help to take new enterprises through their teething stages. They also advocate certain institutional arrangements which are intended to remove difficulties encountered in earlier times. One difficulty to which reference has been made is that many workers' co-operatives began operations without reasonable market prospects and expecting to make use of an obsolete technology while competing in a hostile competitive environment. The feasibility of the proposed enterprise has to be established before any other steps are taken. A UNIDO Expert Group advises that a feasibility study, which treats all relevant aspects (technology, manpower, finance, market etc.) be undertaken before the formation of the co-operative.[13] It may be preferable to 'pilot' the projected co-operative business simply as a partnership initially until it is seen to be viable and can be registered under the Industrial and Provident Societies Act.

Finding sufficient capital was a difficulty the early workers' co-operatives encountered. Investment in plant and machinery

---

* Co-operative Productive Federation Ltd., 42 Western Road, Leicester LE3 0GL.

left too little to finance necessary purchases of raw materials, stocks and work in progress as well as paying wages and incidental expenses. A device which was often resorted to was to invite trade unions and retail distributive co-operative societies and occasionally local authorities to join together with the workers as shareholders. Shares in a workers' co-operative society remain at their nominal value and are transferable but the transfer requires approval by the committee of the society. Workers in the society can build up their initial minimum shareholding through their allocation of the share in surplus. Co-operatives operate the rule of 'one man one vote' which means that a large shareholder is entitled to only one vote, while the shareholder with a minimum holding has equal voting power. There is generally a ceiling placed on the interest payable on loans taken by the societies.

The disposal of surpluses generated in trading by workers' co-operatives is governed by quite detailed rules. Generally there is provision for appropriation of surplus to share capital holders (as interest and dividend) to employees (in proportion to their salaries and wages), to customers as dividend (in proportion to their trade) and then to various funds (e.g. education, provident, special services) as well as to reserves. The rules can be very specific as to the way the appropriation is made, e.g. 'For every penny on his trade that a trading member receives the worker shall receive 2d. in the £ on his wages and salaries.'

The democratic machinery of a workers' co-operative is particularly important. The Board of Management of the Society is elected by the members. Some societies make a specific requirement as to the number of members of the Board who are to be employees. The Board of Management appoints the officials of the Society. It is the Board of Management that appoints the Chief executive of the Society upon whom, as has already been seen, so much depends. Realising the importance of this appointment, as well as that of other specialists, the CPF offers a staff Selection Service. It is in the area of centrally provided services of this kind, which give specific recognition of the special character of the constituent members' problems that the value of the Federation is to be found. The wider experience of Board Members not employed in the co-operative themselves can prove a useful source of judgement in making appointments or in management more generally.

CHARACTERISTICS

From the foregoing account of the institutional arrangements adopted in this country it will be evident that there is an important distinction between co-partnership co-operative societies and simpler workers' co-operatives. A workers' co-operative is characterised by:

(1) Wage-earners forming an association which undertakes the entrepreneurial function;

(2) The democratic conduct of the affairs of the business.

So in this workers' co-operative the risk-bearing, the mobilisation of capital and other factors of production is undertaken by workers who agree *amongst themselves* as to how their affairs shall be conducted and by whom. In the co-partnership because capital is found from an external source there has to be agreement beyond the workers' themselves on decisions affecting the conduct of the enterprise. Hall and Watkins argue that the co-operative co-partnership '. . . corrects the instability of the self-governing workshop by introducing a body of shareholders who will support the authority of the management'.[14] In a different context Watkins argues that in the co-operative co-partnership it is possible to achieve a synthesis of the useful elements in syndicalism and collectivism. He writes:

> The synthesis becomes possible because industrial management is not a simple but a complex thing, and the variety of functions it denotes makes possible a diffusion of responsibility. The formula of *collective ownership and co-operative organisation* indicates the line of division between the functions of the community of consumers on the one hand, and the specialised group of producers on the other (my emphasis).[15]

The dichotomy that Beatrice Potter saw between 'individualist associations of producers' and consumer co-operatives begins to appear less fearsome, because it is possible to effect a merging of interests.

The practical interest that consumer's co-operatives took in workers' co-operatives stemmed in part from the aim which the earliest distributive co-operatives had of providing employment for their members but also from the desire to invest surplus funds in enterprises from whom they purchased and with whose outlook they agreed. As so often happens the practical requirements of the situation had an outcome with important theoretical implications and provided a solution to problems which had been dogging others. When the trading link becomes weaker, as it has tended to become in recent years, then the solution of balancing consumer and producer interests within the same organisation through a recognition of areas where legitimate interests begin and end begins to look less convincing. There may be other institutions whose role might replace that of the consumers' co-operatives.

DOES IT WORK?

If the figures for 'dividend to labour' are used as a measure of the success of existing workers' co-operatives then one might conclude that the performance of the British Societies was unimpressive because few societies pay a dividend to labour. But this would be to mistake the character of the experiment. As in any other enterprise there are multiple objectives. The CPF describes a co-partnership as:

> . . . a form of group self-employment designed to provide maximum security of employment and income for the individual worker. It encourages joint responsibility for productivity and efficiency in production and management, and offers economic freedom through co-operative labour association.[16]

Some of the objectives are more important than others in setting the value system within the enterprise. One view is that: 'The essence of co-partnership is not profit-sharing but the distribution of rights and responsibilities.'[17] If this is so then one would expect the society to show superiority in terms of 'growth' or 'efficiency' or stability in employment or morale or personal attitudes amongst the workers if not in terms of the division of the surplus enjoyed by the workers. Unfortunately knowledge of the day-to-day effects of working in a co-operative is not available. The UNIDO report already referred to suggests that there is room for 'a series of detailed studies of motivation, discipline, team-work, personal satisfaction and other factors in workshop performance in co-operatives as compared with other enterprises.' This would certainly be a worthwhile international study. Without comparative independent studies we are left with statements like those in the UNIDO report that the strength of producer co-operatives '. . . lies in good factory morale, which invariably prevails in a period when discontent, frustration and conflict are only too frequent among industrial workers.' Or, more generally, that despite the absence of rigorous psychological studies:

> There is a very general impression that the level of social satisfaction among workers in a self-governing co-operative workshop or co-partnership is higher than in other types of enterprises.

The indicators are the length of service, the absence of dismissals, strikes or industrial disputes, involvement in the formal democracy of the co-operative, the absence of tensions and a good 'atmosphere' in the enterprise.

On the formal aspects of democracy within workers' co-operatives rather more is known. It is possible to measure, if crudely, the involvement of workers in the democratic machinery of the societies. What remains unknown is the extent of their influence in areas of management decision. The results of a survey by Geoffrey Ostergaard showed an average attendance of 10.2% of the membership at business meetings of co-partnership societies.[18] The attendance ranged from 6.9% in clothing societies to 15.3% in the printing societies. An even higher rate of attendance (16.3%) was shown in a miscellaneous group of societies some of which were of recent formation. Attendance was highest in societies with fewer than a hundred members (38.9%) and fell away markedly in the societies with over a thousand members (4.6%). Perhaps the most interesting aspect of Ostergaard's survey is his tabulation of member attendance by the category of membership. He shows that workers accounted for 67.5% of all those present at the meetings. Individual shareholders were next in importance (21.7%) while co-operative societies (9.9%) and trade unions (0.9%) were less in evidence. The turn-out of worker-shareholders represented 25.9% of those entitled to attend, a figure close to that for trade unions (25%) while co-operative societies and individuals showed a much lower turnout (3.9 and 6.3% respectively). Ostergaard concludes that: 'Clearly the degree of effective worker control in co-partnerships is greater than might be expected from the membership distribution figures.' This finding is one to which I shall revert again later.

CURRENT POSITION IN BRITAIN

In 1971 there were twenty-three active workers' co-operatives registered under the Industrial and Provident Societies Acts in England and Wales. The sales of these societies in 1970 (the last year for which a complete set of returns was available) was £4,877,000.* In 1968 the sales of the same societies amounted to £4,543,000. These societies employed 2,265 workers in 1970, a somewhat lower figure than in 1968 when it was 2,453. The most significant group of societies is in the clothing trades where there are four societies with 40% of the total combined sales and 48% of the total employment. Four footwear societies produced 28% of the workers' co-operatives combined sales, while eleven printing societies are almost as important with 20% of sales. The remaining societies are in engineering, vehicle building, professional services and house building. Virtually all the co-operatives can be regarded as 'small businesses' although one society (Ideal Clothiers Ltd.) have sales in excess of £1 million and close to 600 employees. In 1965 there were 40 societies in existence. The growth in sales of these societies over the period 1968–71 was close to 2% p.a. for the footwear societies, 3% p.a. for clothing and 4% p.a. for printing. (Some of the sales figures for 1971 are estimated.) This does not suggest a high rate of growth in an inflationary situation. Some have been absorbed by the Co-operative Wholesale Society while others have been wound up. The Co-operative

---

* Author's estimates.

Union's Annual Statistics for all Productive Societies (including those in liquidation) showed 26 societies in existence in 1970 against 40 in 1965 (the statistics include England, Wales and Scotland). In 1965 there were nine societies in the clothing and allied trades, while in 1970 statistics show five. In the other main categories printing showed no change in numbers over the period, while the footwear and allied societies declined from ten to four over the five-year period. The Co-operative Union's total sales figures for 1965 were £7,198,000 and for 1970 £5,620,000.

EUROPEAN EXPERIENCE

The strength of workers' co-operation in Britain can be contrasted with that in two other countries in the EEC, France and Italy. In the mid-1960s there were about 550 workers' co-operatives in France with some 35,000 members and a turnover of £55 millions (at the exchange rate then current). The French societies do not follow the British pattern but maintain the tradition of ownership and control solely by the workers. There are important differences in the distribution of employment too. In France there were, in the 1960s, about 350 co-operatives in building and construction, competing for local and central government contracts as well as contracts for private clients. The public authorities in France are asked to assign contracts in such a way as to assist the co-operatives.* L'Hirondelle, the largest of the Construction societies, which was formed in 1920, built the Hilton Hotel at Orly Airport. Amongst the remaining 200 societies there were significant groupings in paper and printing, engineering, woodworking, leather and textiles, food and glass and pottery. The 'professional services' category is quite developed including accountancy, town planners, architects and surveyors. Forestry work has attracted some co-operatives in France. In the electronics

---

* The disaster which overtook the National Building Guild in Britain in 1921 when the Government altered its policy for pricing contracts may be recalled as a contrast to the French experience.

industry ADIP makes precision instruments (some for use in nuclear plants), equipment for telephone exchanges and navigational equipment. ETCM has won a considerable reputation for prefabricated farm silos. The 'Familistere' a society founded by a follower of Fourier produces heating equipment, metal furniture and refrigeration equipment.

The picture in Italy is just as diverse. There are a number of large building and construction co-operatives. The statistical base does not appear as firm in Italy but something in the region of 3,500–4,000 co-operatives exist with about 125,000 members. A variety of societies produce amongst other items optical and dental equipment, weighing machines, lifts, mining equipment, machine tools, grain driers, and fruit-grading equipment. The textile clothing and shoemaking trades have their co-operatives. Transport co-operatives are in evidence in Italy (as in France too) and engage in road, rail and sea operations. In Italy co-operatives have been formed to provide services of various kinds, particularly in tourism. Security services, firewatching, and diving services are all provided by workers co-operatives.

The French experience of workers' co-operation is an impressive one with growth rates better than national averages and investment outpacing that in private industry. An observer well-placed to assess the reasons for this favourable development in France stresses that:

> The real economic progress of the French productive Co-operatives started around 1950 at the same time that a methodical policy of education among the members began to be carried out. The most successful co-operatives in the economic field are also the ones where control by the working members is the best organised ... Our experience in recent years has taught us that a separate co-operative even when very strong is unable to carry on all the educational duties that go with real control by the members if left alone with its own means. Accordingly, the part played by the Federation* inside the Co-operatives

---

* *Confédération Générale des Sociétés Coopératives Ouvrières de France.* (CGSCOP)

in the successful organisation of information and educational programmes is bound to widen and strengthen every year. This goes hand in hand with the permanent and ever-increasing work of economic, commercial and management assistance.[19]

The problems which arise in recruiting top rate managerial and specialist staff and integrating them in the workers' co-operative which requires their skills if it is to survive and grow is a crucial one. Coming to terms with this problem has helped to highlight the educational problem and an interesting diagnostic tool devised by CGSCOP is the Bilan Co-operatif, a 'social' balance sheet which provides various indicators of the democratic, manpower resource and morale status of the co-operative. Through its emphasis on transmitting information to workers and carrying through a programme of co-operative, economic and technical training the French workers' co-operatives have begun to make the decentralisation of decision-making more real to everyone employed in their establishments.

CONTRASTS IN EXPERIENCE

The vitality of French workers' co-operatives owes a good deal to the fact that they could not look for support amongst the French trade unions which were split among religious and political lines or the French consumer co-operatives which only began to emerge with any importance in the 1950s. The workers' co-operatives were isolated and hence developed their own common institutions like the Federation and a Central Co-operative Credit Bank which meant that technical advice and finance were available to its members. Unfortunately the British movement came to be regarded after 1930 by the consumer co-operatives, who took its products and 'shared' in its management as, 'a somewhat doubtful form of co-operation' and 'an appendix which could be removed without loss'. The malaise which affected the co-partnerships in Britain after the post-war economic bubble burst must have its origin at least in part in the failure of the consumers' co-operatives to recognise that the co-partnerships production was complementary to rather than competitive with the products of the Co-operative Wholesale Societies which they owned. As the volume of production which the retail societies took from the co-partnerships has declined they have looked elsewhere for trade and found new customers, both at home and overseas. Their independence has increased and in the meantime many of the productive activities of the Wholesale Societies have been closed down as these organisations concentrate their resources in retailing and wholesaling operations and use their market power as specification buyers. This re-allocation of resources within the British Co-operative movement may have produced a healthier situation so far as the co-partnerships are concerned. As the analysis of their democratic machinery showed the involvement of the retail co-operative societies in the affairs of the co-partnerships was not particularly pronounced. Perhaps the co-partnership solution will be seen as a 'sport' in the development of workers' co-operation. Possibly the obsession with failure amongst proponents of workers' co-operation led them too readily to assume that tripartite management (workers, trade unions and consumers' co-operatives) was a better solution than the independent society. Small businesses have high failure rates and those formed because of the pressing need to create work for the unemployed who set them up, without a breathing space in which to test their feasibility, are especially susceptible to failure. Given these beginnings and that of all co-operatives:

> The workers' productive co-operative is regarded as one of the most difficult to organise and maintain, since it is concerned not with an outlying activity of its members, but with one that is central to their whole lives and occupies most of their time and thoughts. The contact is with almost the whole surface of their personality, and acute problems of mutual adjustment and avoidance of friction are thus added to those of business management and control of valuable property. (UNIDO Report)

Many experiments are foredoomed. But they are still invaluable training grounds in industrial democracy and to be encouraged alongside the more 'programmed' ventures, to the prospects for which I next turn. The commitment of 'risk capital' in these experiments can be justified in terms of social learning as well as in the more conventional measures of return on capital and profitability. The favourable trade-off for society between investment in new socio-economic institutions and conventional economic returns from more stable forms of organisation has been underestimated. It is unfortunate that the history of these experiments is so poorly documented, their experience should be 'caught' before all traces are lost. Many workers' co-operatives probably never reach the stage of formal registration as societies and it would be pertinent to know what barriers and problems they encountered before dissolving themselves. Sometimes what might have been a workers' co-operative becomes a company because there is a lack of knowledge amongst those who set it up of the advantages that this very flexible form of organisation offers, sometimes the legislation which governs their operation appears too unfamiliar.* We need to know why these potential entrepreneurs did not become established and thereby add to our fund of experience in the management of democratic institutions.

PROSPECTS

If previous developments in workers' co-operation in the UK were extrapolated we might expect societies to remain restricted to a relatively small number of industries where the economics of large-scale production have not yet penetrated and capital investment is less important than skilled labour in producing the final product. These industries could also be those where the convenience of a local service outweighs any cost reduction that might be found in a more nationally-based concern. The limitation of the field of activity would be such as to make the prospect appear a rather gloomy one. There are alternative futures, however, which I want to outline. Suppose that in the 'post-industrial' society:

(1) Basic industry is less important as an income and employment generator;

---

* Freelance Programmers Ltd. would seem to have been in a situation of this sort. *Guardian*, 8 September 1972.

(2) The services sector grows;

(3) A non-profit, quasi-governmental sector emerges 'between' the public and private sectors.[20]

An economy like that described would certainly prove one in which there would be scope for the development of workers' co-operatives. The provision of services, many of them quite new, to consumers and producers would permit an enterprising group of workers to 'stake a claim'. The expansion of the services sector is already an acknowledged economic fact. An American study of producer services lists the following producer services as showing high employment growth rates in the period 1958–63: direct mail advertising; duplicating; copying; services to dwellings and other buildings; research and development laboratories; testing laboratories; business management consulting; detective agencies; equipment rental; coin-operated machine rental, repair; photofinishing laboratories; interior decorating; sign painting; auctioneers' establishments; telephone answering; water softening; supplying temporary help for business.[21] If industrial firms take the view that they are in business 'to make products not wash windows' then there is certainly likely to be a growth in services. The flexibility which comes from the ability to sub-contract may also result in more activities being pushed out of the production unit. Lasuen has described this development as reaching the stage where if possible firms: '. . . would subcontract out most of their activities and would seek to retain exclusive management over the assembly of all the pieces. They would give away the research, procurement, production, selling, servicing and so on, and keep the designing, planning, marketing and financing functions.'[22] One might question in fact if they need retain as many functions as Lasuen suggests they might!

The proliferation of consumer services is more likely to arise from the complex needs of a highly-organised society dependent on more and more technology in the home and the increased demands of leisure time pursuits. These developments and the growth in demand for individualised products for personal use will certainly provide openings for workers' co-operatives.

The future then is perhaps not bleak but full of opportunities. Are there going to be the people prepared to join together to take these opportunities? Again there are some promising developments which suggest some optimism is justified in viewing the future. Firstly, there is a general recognition that, to quote Daniel Bell, 'The most characteristic fact about the factory worker today is his loss of interest in work. Few individuals think of "the job" as a place to seek any fulfilment.' The recognition of this fact must incline management and workers towards experiment and the co-operative should be amongst the alternative forms considered since it has, at least, been tested both in this country and abroad for more than a century. Secondly, the movement away from autocratic towards participatory management styles should foster those habits of thinking which will incline workers in industry towards a truly democratic participation in decision-making in enterprises they own themselves. The outcome of both these trends must surely be that more workers will recognise that the present form of industrial organisation is inadequate '. . . because it restricts the simple wage-earner to a role altogether too passive in the organisation of his work, because it requires from him not conscience but only obedience.'[23] There is still the crucial issue of management for the new co-operatives to face. One possibility which has been advanced and should not be ruled out is that:

> Perhaps the changing national attitudes today will throw up a new group of skilled managers who, disillusioned with the destructive conflicts of an authoritarian company, will choose to work in a productive society.[24]

Certainly the existing co-operatives are too few to develop the necessary management themselves. The shift in attitudes which has been brought about by disillusion with conflict and the spread through management education of knowledge of alternative styles of management will, however, facilitate finding and training managers for new workers' co-operatives once they are formed. But how are workers to learn that there is an alternative form of industrial organisation? Their own trade unions have dismissed the experiment as 'outmoded' and the consumers' co-operatives no longer show an interest in the ideal of creating co-operative employment even through co-partner-ships. Is the dissemination of knowledge about this social innovation to be left to the few existing co-partnerships or is this a function which society itself should assume through an appropriate body? Is it enough to have a Registrar for Industrial and Provident Societies? Might not the way progress amongst Housing Co-operatives was stimulated provide a model for the future? It is to be hoped that a future government will decide that it can 'trust the people as individuals as well as in the mass'[25] and stimulate thinking and support action amongst workers to start their own co-operatives.[26] The benefits in terms of finding a solution to problems which destroy the quality of working life could be immense in comparison to the costs of investment in spreading information and supporting the experiment.

REFERENCES

1. This quotation, like those before it is taken from Benjamin Jones, *Co-operative Production*, 1894. This despite its imperfections remains the standard source on workers' co-operatives in the later nineteenth century.
2. The description of Greening's thinking follows J. J. Worley's 'The Producers' Theory of Co-operation and its Relation to the Development of Co-operation', *The Co-operators' Yearbook*, 1929.
3. Neil J. Smelser, *Social Change in the Industrial Revolution*, Routledge and Kegan Paul, London, 1959.
4. Torben Christensen, *Origin and History of Christian Socialism*, Aarhus, 1962. This is the most perceptive account of the Christian Socialists yet to appear. An older book, published in 1920 by C. E. Raven, *Christian Socialism*, remains a valuable shorter study.
5. G. D. H. Cole, *A History of Socialist Thought: The Forerunners, 1789–1850*, Macmillan, London, 1953, contains a useful résumé of the ideas of the early French and British Socialists.
6. Amos Mann, *Democracy in Industry*, Leicester, 1914.

7. H. Clegg, *A New Approach to Industrial Democracy*, Blackwell, Oxford, 1960.
8. P. Blumberg, *Industrial Democracy: The Sociology of Participation*, Constable, London, 1968.
9. D. Robertson and D. Dennison, *The Control of Industry*, Nisbet, London, 1960.
10. This passage is taken from a paper, 'The Relationship Between Co-operation and Trade Unionism' which appears in *Problems of Modern Industry*, Sidney and Beatrice Webb, 1920. A considerable part of the *Co-operative Movement in Great Britain* which she published in 1891 as Beatrice Potter is devoted to an attack on Association of Producers.
11. The Report appears in G. D. H. Cole and A. W. Filson, *British Working Class Movements: Select Documents 1789–1875*, Macmillan, London, 1951.
12. These remarks are taken from Alfred Marshall's *Elements of Economics of Industry*, Macmillan, Sixth Edition, 1899.
13. United Nations Industrial Development Organisation, *Nature and Role of Industrial Co-operatives in Industrial Development*, Vienna, 1969, Ref. ID/25.
14. F. Hall and W. P. Watkins, *Co-operation*, London, 1937.
15. W. P. Watkins, 'The Co-operative Ideal in Industry', *The Co-operators' Yearbook*, 1927.
16. *Co-operative Co-partnership Conspectus*, CPF, 1971.
17. Watkins, 1927, op. cit.
18. G. Ostergaard, 'Member Participation in Co-operative Co-partnerships', *The Co-operators' Year Book*, 1957.
19. Antoine Antoni, 'Workers' Control in French Co-operative Societies', *Review of International Co-operation*, Vol. 61, no. 6, 1968.
20. This draws on C. H. Madden, 'Business and the Future', *Futurist*, December 1969.
21. Harry I. Greenfield, *Manpower and the Growth of Producer Services*, Columbia University Press, New York, 1966.
22. J. R. Lasuen, 'On growth Poles', *Urban Studies*, June 1969.
23. Watkins, op. cit.
24. Roger Sawtell, 'Sharing our industrial future?', *Industrial Society*, 1968.
25. G. D. H. Cole, 'Co-operation, Labour of Socialism',

Blandford Memorial Lecture, 1955.

26. The Labour Party's policy document *Labour's Programme for Britain* contains proposals for a Co-operative Development Agency with finance at its disposal for new and existing developments.

The Coal Industry

PETER ANTHONY

Section 46(i) of the Coal Industry Nationalisation Act, 1946, stated that:

'It shall be the duty of the Board to enter into consultation with organisations appearing to them to represent substantial proportions of the persons in the employment of the Board' and to conclude agreements with them providing for the negotiation of terms and conditions of service and for 'consultation on (i) questions relating to the safety, health or welfare of such persons; (ii) the organisation and conduct of the operations in which such persons are employed and other matters of mutual interest to the Board and such persons . . .'.

The particular form of consultation was left to the Board to decide. The general duty to consult and the form which consultation took were not very different from the proposals which had been made by the coalowners to the Sankey Commission some thirty years before. The fact that the coalowners were prepared to establish a scheme of consultation at a time when the whole weight of mineworkers' representative opinion was committed, not merely to nationalisation, but to joint control and direction of the industry, illustrates their difference of outlook: it may also illustrate the speed and direction in which labour thinking had developed in the intervening years.

The phrase 'other matters of mutual interest' was open to a variety of different interpretations; what was and what was not a proper subject for consultation was often a matter for argument within the consultative channels. The Board's perception of the purpose to be achieved by joint consultation began to appear in the early Annual Reports and Accounts following nationalisation. This purpose was twofold: to enable management to secure the experience and knowledge of workers in helping to solve the problems of the industry and to promote unity. The achievement of a sense of unity among employees must have seemed particularly important. By 1948 the Board was beginning to describe joint consultation in terms appropriate to an evangelical movement: 'Faith in consultation among management and men is spreading.' The Board stressed faith and good works—faith in that the 'success of consultation depends not so much on the network of committees . . . as on the attitude of mind . . . The Board believe that there is now within the industry the will to make joint consultation work'; good works in that the Board constantly publicised examples of the right way to practise joint consultation. Some measure of its importance can be found in the Board's Annual Report for 1948 in which eight pages out of a total of 150 are devoted entirely to joint consultation.

There were some signs, however, of faith being challenged by doubts and of its eventual decline. In 1952 a perceptive comment appeared in the Annual Report on a problem which was to bedevil joint consultation: 'The danger is that the machinery of consultation may continue to work smoothly only because controversial or difficult subjects are avoided . . .' After 1952 the subject of joint consultation begins to drop out of the Reports and the special section devoted to joint consultation disappears.

A new section appeared in 1962, significantly headed 'Communications'. The purpose is the same, to create a sense of unity, but the means by which 'a sense of common purpose' is to be achieved now includes a battery of techniques and new agents: 'Coal News', 'Management News', Staff College courses and conferences. There is considerable evidence to suggest that the Board had perceived joint consultation as a managerial instrument to be applied as a means of communication with employees in order to produce cohesion and harmony and that the Board finally judged it to be an ineffective instrument for this purpose. If it was ineffective why did it fail to achieve the objectives for which it was apparently designed by management?

One explanation must certainly concern the very different perceptions of the National Union of Mineworkers. In an industry the labour ideology of which was probably syndicalist rather than Fabian, in which the traditional demand was for worker control rather than for mere nationalisation, it would be surprising if management's concern with communication and unity should have achieved total understanding. The National Union of Mineworkers may have abandoned its pursuit of joint control but it could be forgiven for some scepticism over the extent to which consultation was likely to afford it significant influence on the direction of the industry's affairs. To the extent that it was not sceptical it may have believed that the machinery of National, Divisional, Area and Colliery consultative committees would allow some opportunity for influencing managerial decisions as the machinery was claimed, usually by managers, to bear some relationship to industrial democracy. Very little in its experience of joint consultation could have reinforced this fragile faith.

Some early confusions arose when Board representatives insisted that 'once appointed, a member did not represent any particular association' or section, everyone was to work together in the best interest of the industry. It prompted Arthur Horner to wish that the National Council would 'consider abandoning the principle that members of consultative committees should be consulted with as individuals rather than as members of a trade union'.

Differences over the role of members of committees diminished as time went on, but there was continuing uncertainty over the functions and purposes of the committees themselves.

Minutes of consultative committees show the union frequently pursuing some vestigial element of joint management, seeking information, asking to be 'consulted' about impending changes, attempting to find limited areas in which the committee could exercise executive functions. These efforts were often unsuccessful. The Board, in the early years, often refused to disclose information about the financial results of particular units. A Divisional council first began to give production and manpower figures three years after nationalisation. In the case of one colliery, the last reference to production matters at the consultative committee was that a coal face was being prepared for power loading. This was in January 1966. The colliery closed the following month without any reference to closure in the committee's minutes.

There was a general reluctance to allow the committees any

executive authority except when attempts were made to engage the committee in decisions concerning disciplinary action against persistent absentees. One colliery committee interviewed, or endorsed the dismissal of, 241 men between 1951 and 1963. There were also sporadic signs of union resistance to become involved in disciplinary decisions.

In some respects the committees' activities seem almost to be independent of the events taking place in the industry. This impression of unreality is confirmed by the evidence of a detailed analysis of the minutes of four committees (National, Divisional, Area and Colliery) over a period extending from 1947 to 1965. Each item of discussion (almost 10,000 were recorded) was analysed in terms of its subject, the party introducing it, the party's intention in raising it, the conclusion of the discussion, and the degree of conflict or agreement manifested during the discussion.

Figures of coal production achieved were a very marked managerial preoccupation. At the colliery committee, production records represented more than 30 per cent of all the subjects raised by management. In terms of subjects raised, management dominated the committees and this domination increased with time so that, on the three lower committees the percentage of items raised by management had doubled; in two of these committees management came almost to monopolise the meetings. At the national level the movement was in the opposite direction until, in 1965, the NUM initiated 35 per cent of the subjects discussed.

The tendency of the committees to discuss matters of recorded information rather than of change also became more marked. On the Divisional Committee management's initiation of production records reached 41 per cent in 1965. Management's attention to change, at least as represented by the frequency with which it initiated its discussion, was never high and it gradually diminished.

The National Committee was again exceptional. The subjects initiated by management were more evenly spread. Broadly speaking, the National Committee ended the period as it had begun it, while the lower committees became dominated by management talking more often about fewer subjects.

Table 1 shows the way in which the business of the committees was distributed between the four representative groups in terms of their apparent intentions in raising items. The Table gives the items raised under each heading as a percentage of the total items raised in all four committees.

TABLE 1

	Information Management-Employees	Information Employees-Management	Appeal	Complaint	Proposal	Other
Management	55.9	.1	3.8	.1	1.6	.8
NUM	.3	1.6		12.5	2.9	.6
NACODS	.1	.1		1.4	.3	.2
Mgt. Un.	.2	.2		1.4	.4	.2

The fact that management distributed its control of the committee's business by passing information downwards to employees is illustrated even more clearly in Table 2 which shows the distribution of items under each heading as a percentage of the items raised by each group in all the committees.

TABLE 2

	Information Management-Employees	Information Employees-Management	Appeal	Complaint	Proposal	Other
Management	89.7	.1	6.0	.1	2.6	1.3
NUM	1.8	8.8		70.6	15.6	3.1
NACODS	5.2	10.4	.5	60.4	14.1	9.9
Mgt. Un.	9.5	7.7	.9	58.8	15.8	7.2

Almost 90 per cent of the items raised by management were raised for the purpose of using the committees as a channel of communication to employees. The second most significant intention was in making appeals for co-operation. The NUM raised 15.6 per cent of its items as proposals for action (as against only 2.6 per cent for management).

The most significant results emerged from an analysis of the degree of conflict or agreement surrounding the committees' discussions. In committees designed in part to facilitate the meeting of different groups and the reconciliation of different interests there was an astonishingly low level of conflict. In the discussions of the four committees during the period, there was no evidence of either conflict or agreement in 93 per cent of the discussions. The distribution of the items between the various categories of agreement/conflict are shown in Table 3.

TABLE 3

	Category 1	Category 2	Category 3	Category 4	Category 5	Category 6
	Unanimous Agreement	Approval	No Conflict or Agreement	Technical Conflict	Criticism of Individuals or Groups	Value Conflict
All Subjects	.3	1.7	93.1	3.8	.6	.2

The subjects which gave rise both to some degree of conflict and to positive agreement were those concerning change (production plans, manpower plans) or the general condition of the industry. The general tendency was for discussions to concern the statistical records of performance which occasioned neither positive agreement nor conflict.

Table 4 shows the number of occasions on which some degree of conflict was expressed as a percentage of the total number of occasions on which each subject was discussed at each committee.*

This shows a tendency for the incidence of conflict to reduce very considerably from the highest to the lowest level committee: on the National Committee, 11 per cent of the discussion occasioned some degree of conflict while this level is reduced by two-thirds at the Divisional Committee and is halved again at Area level.

* Discussions were allocated to the following subjects:
(1) Production Records, (2) Production Changes, (3) Situation of Industry, (4) Manpower Records, (5) Manpower Changes, (6) Safety and Health, (6) Training Records, (8) Training Changes, (9) Joint Consultation, (10) General.

TABLE 4

Subject	Conflict			
	NCC	DCC	ACC	CCC
	%	%	%	%
1	7	4	3	3
2	23	6	5	5
3	10	8	1	2
4	10	5	2	2
5	14	19	5	7
6	9	3	1	1
7	11	2	3	—
8	6	2	5	14
9	19	4	1	6
10	8	5	3	7
Total	11%	4%	2%	3%

Table 5 illustrates differences in the behaviour of the representative groups in terms of their performance in the four committees. The figures show the items raised by each group under their conflict/agreement category as a percentage of the total items raised by the group.

TABLE 5

	1				2				3				4				5				6			
	NCC	DCC	ACC	CCC	NCC	DCC	ACC	CCC	NCC	DCC	ACC	CCC	NCC	DCC	ACC	CCC	NCC	DCC	ACC	CCC	NCC	DCC	ACC	CCC
Mgt.	1	—	.5	—	4	2	1	1	90	96	97	97	5	2	1	2	1	—	.5	—	—	—	—	—
NUM	2	1	—	—	1	3	4	.5	75	78	88	94	17	12	5	4	4	3	2	1	1	2	1	—
NACODS	1	1	—	—	4	8	—	—	75	82	100	—	18	8	—	—	2	1	—	—	—	—	—	—
Mgt. Union	2	—	—	—	6	—	—	—	76	93	87	—	13	6	13	—	3	1	—	—	—	—	—	—

Almost the whole of the items raised by management resulted neither in conflict nor agreement, but in neutrality. The proportion of management items recorded as neutral (category 3) ranges from 90 to 97 per cent of the total, a higher level generally than any other group (except the NUM at Colliery level and NACODS at Area). The NUM's items are more evenly spread. But whereas management's non-neutral items tended to approval, the NUM's were more likely to concern conflict. The NUM's conflict-producing items also included technical conflict (Category 4), criticism of individuals or groups (category 5) and value conflict (category 6)—the NUM was in fact the only body to engage in this, the most severe, category of conflict.

The percentage of neutral items rises as we descend the hierarchy of committees and this is true for each of the representative groups. On the National Committee, 90 per cent of the items raised by management were neutral, on the Divisional Committee 96 per cent, and, on Area and Colliery Committees, 97 per cent. As the percentage of neutral items rises the percentage of items containing conflict falls. A considerable proportion of the items raised by the NUM and by NACODS on the National Committee (18 and 18 respectively) fell in category 4, technical conflict. These percentages rapidly diminish until only 4 per cent falls in category 4 for the NUM at Colliery level.

A further analysis of the committees year by year shows some degree of consistency for the National Committee. The percentage of items raised by management and falling in the neutral category never rose above 57 (in 1954 and 1956) and was falling towards the close of the period. The NUM's items were fairly evenly distributed throughout the period.

In the Divisional Committee, on the other hand, the percentage of management's items in category 3 rose steadily from 35 (in 1947) to a peak of 81 in 1963 and 80 in 1964. In the Area Committee there were no issues raised by the NUM which occasioned any conflict between 1951 and 1960 and only 2 were raised by management. In the Colliery Committee there were no issues raised by the NUM in category 6 (value conflict) during the whole of the period of the study (that is, during the whole life of the Colliery under nationalisation), and there were no criticisms of individuals or groups after 1958. There were no instances of conflict at all in the Colliery Committee during the last year of the study (1965), or in the first three months of 1966. This is a most singular phenomenon as the Colliery closed early in 1966 and the closure was resisted both by the NUM and the colliery lodge.

The immediate conclusion must be of a massive lack of either conflict or agreement in the minutes of the three lower committees. What changes there are in the distribution of items within the six categories seem to have been evolutionary and could not easily be related to changes taking place within the industry or outside it. The committees do not seem to have responded to events, nor did the representative bodies reflect events which affected their own interests. During the period there were considerable changes in the financial, social and marketing fortunes of the industry. At the organisational levels corresponding to the four committees there were events of major significance which must have affected employees; the period extends, for example, from the predominance of hand-got mining to the successful introduction of power-loading. Within the colliery there were several instances of mechanical innovations which were unsuccessful. Considerable changes were brought about by statute and by management in training arrangements (training, in fact, seems to have been a subject of little interest to anyone on the committees). There were vast changes concerning manpower deployment during the period, including extensive schemes for inter-divisional transfer and two largely abortive attempts to introduce foreign mineworkers. The marketing and commercial situation of the Board underwent violent change. Drastic colliery closure programmes were announced by the Board after 1960. A government White Paper (in 1963) revolutionised the Board's management of its financial affairs.

But it would be impossible to relate the committee's discussions to any of these events. The extent to which the discussions reflect any consequential disagreement develops or atrophies independently of events in the real world of coal-mining. The only reasonable conclusion would seem to be that the committees (with the possible exception of the National Committee) have a life of their own, not related to the expression of interest or to the resolution of difficulties.

We have not yet suggested an explanation for this very low level of conflict or of 'reality' in the discussion. Generally, the explanation lies within Dahrendorf's proposition that joint consultation is a conflict-suppressing technique. More specifically, it is explained by the proposition that the first axiom governing management's behaviour in relation to the Consultative committees is that it promotes the cohesion of the committee which it controls. But why did the other bodies permit this situation to develop? It seems most likely that the low level of conflict initiated by the union results from the union's perception that the committees were not suitable vehicles for the pursuit of its own goals. Even in those subjects appropriate to the consultative channel (safety, health, welfare, training) the evidence suggests that the union (or the union and management between them) constructed a third channel, that of private discussion. H. A. Clegg has referred to this process: 'When the Coal Board has to take a decision which will affect the NUM its first thought has been: "Clear it with Arthur Horner . . ." and not: "Refer it to the National Consultative Council".'[1]

We have already observed that the committees' activities seem to be almost entirely independent of events taking place outside them. It is unthinkable that a Colliery should have been closed, for example, without any discussions being held between the Board and the NUM, but the discussions did not take place in the Consultative Committee.

In 1955 the minutes of the National Council record a reference by the chairman to the publication of the Fleck Report; the chairman outlined the decisions which the Board had taken following the recommendations, but there was apparently no discussion in the Council. In the Divisional Council, the chairman, reporting an appeal by Sir Herbert Holdsworth (chairman) for higher production, added that the Divisional Board would be inviting the organisations represented on the Council to discuss the situation with the Board. This decision established a pattern which continued for many years in which the Divisional Board held a series of annual meetings with each of the unions and associations separately in order to discuss the coalfield situations. The Consultative Committees were often, at best, informed of discussions and meetings that were taking place elsewhere. In 1963 the Divisional council was told by the NUM that it would be sending a scheme for improved safety to the Board and to NACODS and that 'it was intended to invite the co-operation of the Board in certain decisions taken'. 'Private' arrangements to meet also took place at Colliery level. The manager of the Colliery in the study was asked about a problem, he replied that 'he preferred not to discuss the matter as he had already arranged to meet lodge officials'.

The evidence represents only a fraction of the extent to which the consultative channel was deliberately by-passed by the Board and the other bodies. In none of the committees examined were the discussions concerned with the solution of the many complex problems which arose from time to time. Closure, redeployment, training, recruitment, were regarded as events to be reported upon rather than problems which should be solved. Purposive discussions inevitably took place, sometimes within the framework of collective bargaining, sometimes in private meetings, always outside the Consultative Committees.

The problem almost becomes one of explaining that small degree of conflict in which the union did engage on the Consultative Committees. Here we would suggest that the union's selection, both of the Consultative Committees as channels for conflict expression, and of the issues to express it are influenced by a second axiom: the union's choice of the consultative channel and of conflict issues is influenced by the union's perceived ability to recruit other associations to its side. This usage depends upon the accidental characteristics of the way in which committee minutes were distributed and upon the semi-public nature of the Consultative Committees. Conflict within the committees engaged in by the union against management will tend to promote the cohesion of management and of the union, will threaten the cohesion of the committee and might recruit the power of management and of the union at a higher organisational level. But these results are only possible if the higher levels are informed of the conflict.

In a sense when the union engages in this fractured process, it is doing so in order to give notice of the conflict by means of the minutes. This may be far from a peripheral characteristic, it may be an explanation for the choice by the union of the consultative channel as a vehicle for conflict on those rare occasions when it was so used. This public aspect is one of the characteristics of the consultative channel which differentiates it from other channels available for the expression of conflict.

The argument suggests that the choice of issue to be expressed within the consultative channel will be influenced by whether the issue is such that the union's purpose will be assisted by the attention of higher management. Issues of this kind will be those in which lower management has been acting contrary to higher management's decisions and against the interest of the lower level union. Thus, where management can be accused of impeding production and lowering wage levels by inefficiency in the supply of raw materials, the union will use the consultative channel to protest. Where the union can foresee that higher management's view is likely to coincide entirely with the view of lower management, then there is little point in using the consultative machinery, as the conflict is more likely to be expressed within the conciliation machinery or within private meetings because a necessary prerequisite to the union achieving its purpose is that management's objections must be overcome. It may indeed be easier for the union to do this if it is assured of privacy and protection from the attention of both higher management and other interested parties (hence the choice of the bargaining machinery or the private channel).

We may propose that consultation is a relationship suitable only to the expression of fragmented or stratified conflict (which involves conflict with an external organisation) or intra-organisational conflict (in which hierarchical levels of the organisation are divided against themselves) but that the more usual forms of monolithic conflict (that is, simply, union v. management) must be expressed in other channels.

The more general conclusion, however, is of the absence of conflict rather than the existence of a special form. The most obvious explanation concerns the constitutional exclusion of matters concerning terms and conditions of service. Clegg has argued that after the war, 'the essence of the philosophy was that joint consultation was the means to produce a new industrial society. To this end it emphasised the differences between collective bargaining and joint consultation' and that 'experience of the difficulty of controlling shop stewards was one of the grounds for exclusion of all bargaining issues from the scope of British consultative committees'.[1] But it seems likely that there is a deeper theoretical foundation for the separation.

It seems likely that joint consultation was designed to achieve what Kaplan has distinguished as psychological participation ('the extent of influence on a jointly-made decision which the participant *thinks* he has') as opposed to objective participation ('the amount of influence on the decision which he actually has').[2] If so, it must have seemed necessary to the designers to exclude any issues which would expose the basically calculative relationship existing between employers and workmen, the inclusion of collective bargaining would have allowed the intrusion of a calculative involvement which must have been perceived as threatening the effectiveness of the moral involvement which was the goal. The result of the intrusion could be expected to be 'some waste of power resources through neutralisation, and some loss of involvement because of the ambivalence of lower participants exposed to conflicting expectations (concerning their involvement) associated with the various types of power. Such ambivalence is generated, for example, when lower participants are expected to be calculatively and morally committed at the same time.'[3]

If joint consultation was designed to achieve psychological participation and if the separation of bargaining issues (and the reduction of conflict) was maintained for this end, did it do so?

Many accounts of the failure of consultation have turned on its practical problems. Difficulties of communication have been observed between the 'constituents', the workers and those who represent them on committees. The NIIP stressed the pattern of managerial leadership, technical and economic conditions in the industry, the degree of job security of the worker.[4] Clegg stresses the by-passing of the committees, the absence of an appeals procedure (it seems likely, on the other hand, that the opportunity for fractionalised conflict creates a non-formalised appeals procedure), the limitations on the power of local management and, most seriously, managerial incompetence; managers 'have not known how to run a successful committee and they have not wanted to'.[5]

In terms of this last criticism there is little evidence of managerial hostility to consultation and considerable evidence of attempts to ensure that it was administered with professional efficiency. The Board did what it could to encourage the 'spirit

of consultation' by coalfield conferences, by courses for chairmen and secretaries, by incorporating consultation into industrial relations courses for managers and by frequent publicity. If managers continued to fail because 'they have not known
how to run a successful committee and they have not wanted
to', then we must accept the unlikely conclusion that they were
all wilful, or stupid, or both.

There is an alternative explanation, that managers failed to
make joint consultation work because it was unworkable. There
was no shortage of advice on how joint consultation should be
administered and no deficiency in management's will to improve it. But there was a point beyond which improvement
would not have any perceptible effect on the achievement of
goals at which joint consultation was directed. This is because
joint consultation is dysfunctional; at best it is inappropriate
and wasteful in achieving goals, at worst it is likely to impede
their achievement.

The evidence of the Board's published statements clearly
establishes that joint consultation was directed at the improvement of industrial relationships by providing improved communication between employer and employee which, it was
believed, was necessary and sufficient to bring about a sense of
participation in the direction of the industry's affairs. Our own
analysis suggests the dominant importance given to the communication of information from management to employees.
Studies outside the mining industry have reached the same
conclusion. Rees, for example, noted that although consultation
has 'achieved little in terms of industrial democracy, there has
been a growing awareness on the part of the employers of the
value of joint consultation as a method of consultation'.[6] Clegg
and others have identified Elton Mayo and the human relations
school as influential in emphasising the importance of good
communications and its effect on the building up of teamwork
harmony and co-operation. There is ample evidence in the
National Coal Board's published statements to suggest that the
Board intended joint consultation to bring about a sense of
unity in the industry and that its communication function was
believed to be important in this respect.

But the analysis of minutes suggests that this was an unlikely conclusion. The information which was transmitted in the consultative committees was largely concerned with historical data relating to the past performance of the industry and it was selected to concern subject areas which were likely to be free from conflicts of interest or values. Not only was the choice of the information to be communicated mistaken, but the view that communication is sufficient to achieve a sense of commitment is probably ill-founded. It has been attacked by a variety of distinguished critics of human relations theory.

Dahrendorf warns that 'the attempt to obliterate lines of conflict by ready ideologies of harmony and unity in effect serve to increase rather than decrease the violence of conflict manifestations'.[7] Co-determination, for example, he regards as 'an ill-conceived pattern that contradicts rather than supports a general trend toward the reduction of violence and intensity of industrial conflict'. This conclusion is arrived at by the familiar argument that influential worker representatives entangled in managerial functions are diverted from their original objectives. It can be extended to joint consultation because, although less influential, 'entanglement' and its dire consequences depends not upon the effectiveness with which representatives are associated with the control of an industry, but also upon the extent to which they believe they are so associated. The irony of joint consultation is that while it affords worker representatives no significant influence on the direction of the enterprise, it sets out to deprive them of the influence which they might have exerted from a position of independence. If it succeeds then it is open to the criticism which Dahrendorf levels; if it fails it wastes management's resources and is ineffective.

How effective was joint consultation in the coal industry?

From the workers' point of view it was the last survival of the case for joint control of the industry. Certainly, the case had been captured and disguised by management theorists, but the felony had been approved by trade unionists and politicians; it was, after all, a labour government that incorporated the duty to consult in the Nationalisation Act. Joint consultation, in fact, bore no resemblance to industrial democracy. It afforded workers and their representatives little influence on the direction of affairs. Discussions concerned events that had taken place rather than changes to be made and the low incidence of conflict demonstrates the weakness of worker influence. This is not to say that the coalmining industry was devoid of any characteristics of worker influence or of objective participation; but that the influence was applied through other channels.

The most important channel was probably the collective bargaining machinery which established the negotiating relationship between the NUM and the Board. The incidence of unofficial disputes (about half the total number of all British disputes up to the early 1960s took place in the coal industry and all coalmining disputes were unofficial) suggest that the mineworkers themselves were not without the means of exerting pressure to achieve what they wanted at local or pit level. There is also, as Will Paynter described it, in contrast to formal joint consultation, 'the informal and intimate consultation that has grown up at all levels of management. . . The informal consultation is close and personal . . .' It is conducted 'on the basis of mutual confidence and is a well-developed procedure in the mining industry at all levels which makes it possible for both sides to anticipate events and problems and to budget for them in advance'.[8] Mr. Paynter is well aware of the dangers of workers' representatives becoming involved with the industry's management but he believes that, 'provided the union retains its independence and the union representatives their integrity, it is to the advantage of the work-people.'

The problem is probably more complicated than this, in so far as the union's participation in consultative committees is concerned, because consultative committees may be intended to reduce the union's independence to some extent. The Board set out, in the early days after nationalisation, by insisting that consultative committee members should not represent any particular group but should speak for the industry as a whole. This particular assertion was abandoned in the face of reality but it illustrates the nature of the problem, that while union representatives must continue to see themselves as representing the workers, a section of the industry, the industry's management will seek to recruit them into support for their own direction of the industry as a whole. In these terms, management will see the consultative machinery as a device for facilitating this process of recruitment. The union leaders on the other hand, if they are to heed Mr. Paynter's warnings concerning integrity and independence, must resist this process so that their attitude to joint consultation becomes, at best, ambivalent and apathetic, at worst, hostile.

The coal industry was fortunate in that it had constructed other channels in which the independence of the union could be asserted. The existence of these other channels might have saved the industry from the violent conflict which the suppression of conflict in the consultative channel could have caused. Their existence might also have prevented the union from pressing for the reform and expansion of the formal consultative process: if the union saw the consultative committee as an ineffective means of exerting influence they chose alternative means and abandoned joint consultation to management's control.

There were several instances which illustrate the extent to which joint consultation was perceived by the union to be a managerial instrument. In 1949 the NUM refused to take part in joint consultations at Penrikyber Colliery because they objected to the introduction of machinery without the committee's knowledge. The union withdrew from consultation at Merthyr Vale Colliery for three months in 1952. There was a breakdown in consultation at Treherbert Colliery in 1953. In 1958 a wage dispute at Penrikyber Colliery resulted in a decision by the NUM's Area Executive to withdraw from joint consultation throughout the South Wales coalfield. The union took no part in consultative committees for six months. In 1964 the NUM withdrew on a national scale from a series of consultative conferences organised by the Board.

These withdrawals were a significant indication of the union's attitude to joint consultation. We know of no instance where the union broke off its relationship with the Board within the collective bargaining machinery. This is not surprising. Collective bargaining is, for both management and union, an instrumental and necessary activity. But in terms of joint consultation, the union apparently regarded the *abandonment* of its association with joint consultation as instrumental. The union withdrew from it in order to exert pressure on management. It suggests that the union perceived consultation as a managerial instrument against which the union could apply the sanction of withdrawal.

It is as a managerial instrument that joint consultation must be judged.

Management regarded joint consultation as a means of increasing psychological participation; by means of improved communication it intended to achieve a sense of unity, a commitment to the organisation which would result in a reduction of conflict in the industry. But management confused ends with means; because conflict was intended to be reduced or eliminated by the consultative process its expression was discouraged in that process so that, perhaps, joint consultation could be shown to be 'effective' by the very absence of conflict which characterised it. It was an understandable confusion. Managers were encouraged to believe that consultation would avoid conflict, that conflict was the result of mistaken attitudes and bad communications and that its consequences were usually disastrous. So managers avoided it, particularly in the channels designed to obviate it.

One consequence was that joint consultation could not become a forum for the meeting and adjustment of different interests and attitudes. The industry was characterised by a high level of industrial disputes but there was little dispute in the consultative channels. The industry underwent considerable changes but change was not often discussed by the committees. Many contentious matters were settled between the union and the Board, apart from those relating to wages, but they were settled elsewhere. In terms of one of its main objectives, the promotion of cohesion, the consultative machinery was an irrelevance because it was so much concerned with unity that it could not take the steps to achieve it.

The Board, however, continued for many years to stress the importance of joint consultation as a means of achieving harmony and as a means of tapping the expertise of workers. It would be easy to attack the Board for the inadequacy of its understanding but the attack would not be justified. The Board began its existence with the constraint of a legal requirement to consult. The requirement was the outcome of a complex and shifting political process which had developed over half a century. If the philosophy of joint consultation was confused, the confusion stems from a long series of political adjustments between views which were often inarticulate and rarely consistent.

Faced with the problem of operating under this uncertain brief, the Board sought to be a 'good employer'. It was influenced in its earnest endeavours by the best opinion of the day and made a determined attempt to apply human relations theory to its activities. Goldthorpe has criticised, in terms of the Board's employment and training of its supervisors, 'certain misconceptions in regard to questions of status and conflict which would seem to be fairly common within industry and which may, so it seems, in some cases lead to unsound management policy.'[9] The misconceptions were indeed fairly common and they existed as the result of the influence of academic social theorists whose views were widely publicised and were almost exempt from challenge until the mid-1950s. 'During the war, joint consultation was quite widely welcomed by management writers. In the form of "joint production committees" it also had the full support of the government. There was the expected spate of papers in the journals, especially in *Personnel Management*, welcoming the idea as the start of a new era in industrial relations.'[10] Child quotes C. H. Northcott, G. S. Walpole, Sir G. Cunningham, and R. D. V. Roberts as particularly enthusiastic and influential, but joint consultation had many other sponsors. Managers cannot be blamed for going to what appeared to be the best authorities of the day.

If joint consultation has been a failure, what is to be done about it? If, following Dahrendorf, we agree that joint consultation is built upon misconceptions as to the nature of conflict, if its theoretical foundations are unsound then it cannot be repaired by tinkering with its machinery.

Joint consultation, like anything to do with participation or industrial democracy, is hedged about by ambiguities, some of them deliberately propagated by parties to it whose interest it is intended to promote. Any replacement or improvement of consultation must be directed at the achievement of one of several different, even incompatible goals which were all confused and subsumed as the aim of joint consultation.

The first of these concerns the establishment of effective channels of communication from management to worker. This was certainly one of the aims with which the Board embarked upon joint consultation. But the problems of diffusion and distortion and of loss of signal power are well known in the consultative process, and the Board showed every sign of choosing other and more effective means. It is significant that in Lord Robens's *Ten Year Stint*,[11] while there is no specific discussion of joint consultation and no single reference to it in the index, there are two whole chapters concerned with communication and twelve lines devoted to it in the index. The Board has used every device available to communicate with its workforce; it requires no advice in this field.

Communication, however, is more than a process of disseminating messages by specialised means, it is an integral part of the process of management. The Board has made at least two formidable changes in management structure. It introduced, in 1966, new methods of accountability and control (which bear some relationship to MBO but which may better be described as a process of devolved rolling planning and control). At the same time it reduced its levels of management from five (National, Divisional, Area, Group and Colliery—coinciding, apart from Group, with the consultative levels in our study) to three (National, Area and Colliery). While it is probably true that these changes simplified and improved communications within the managerial organisation, there is nothing to suggest that they improved or were intended to improve the day-to-day process of communication between manager and man. The essential link in this process has always been the deputy and his role (identified as critical in 1960 by Goldthorpe) has always been difficult and may not have been improved.

Communication from management to men may be facilitated or hindered by the organisation but it must depend, in the ultimate, on the state of informal relationships between men and their managers at the point of production. The Board has not relied only on the techniques of modern methods of communication. It has developed and encouraged a process known as 'coalface team conferences' in order to improve relationships and communication. It may well be that developments like this may serve not only to improve communications but to promote that very sense of identification which joint consultation was expected to achieve. There is some evidence of success attending the development of informal, workplace consultation between the manager and his immediate subordinates in the electricity industry. There are inherent advantages in promoting consultation informally at the workplace; it relates to the worker's immediate environment and to real circumstances of which he has direct experience, he is more likely to identify with a work group and with managers engaged in a common task than with an industry which, for him, may be an abstraction. A process of informal consultation at the workplace may also be less subject to the emasculating outlook which has surrounded formal joint consultation and which prevents it from concerning itself with matters of different interest which are likely to provoke conflict; consultation at work, if it is to concern work, will find it less easy to avoid real problems and real interests and, to the extent that it cannot avoid them, it is likely to be more successful in the pursuit of integration in work.

We should always recall, however, that integration is a chimera and that its pursuit, although probably unavoidable by managers, is always likely to fail. Managers have become familiar with Mr. Alan Fox's advocacy of a plural frame of reference which, they are advised, is more realistic than the old unitary outlook which saw all members of the employing organisation as members of the same team committed to achieving the same goal. A plural frame of reference involves the acceptance of the reality of conflict between different groups with different aspirations. In discussing 'New Modes of Joint Regulation', Fox explains that 'only by fully recognising and accepting the constraints imposed by the aspirations of its subordinates, and working through these constraints towards a new synthesis, can management now enjoy any creative role in its handling of the social organization'. He believes that productivity bargaining offers the opportunity of accommodating conflicts to the advantage of the organisation that contains them.[12]

Two qualifications seem to be necessary to the general acclamation which has greeted the notion of pluralism and the techniques of productivity bargaining. Productivity bargaining has been criticised at every level from the ideological to the practical; there would seem to be grave dangers in its uncritical acceptance. Pluralism, while it represents an advance towards realism from the myopic understanding of employee relationships exhibited by some managers, is not devoid of assumptions which can be criticised. We should notice that it is management while 'working through towards a new synthesis', which retains 'a creative role' in 'handling... the social organisation'. Pluralism, in short, is advocated because, being more realistic, it can result in more effective managerial control by way of 'new modes of regulation involving . . . a closer collaborative pattern of relationship'.

Any deceptiveness which accompanies the adoption of a plural frame of reference does not necessarily disqualify it as a recommendation for managerial use, however. To argue that the recognition of the diverse interest groups which constitute an employing organisation is merely a prelude to their more effective control by a more sophisticated management may be a serious matter to every group concerned except the management which, to the extent that the charge is justified, can only gain by the process, however morally suspect it may be.

The principal criticism of joint consultation has been that it is dominated by a unitary frame of reference. What relationship could be substituted for it which would more clearly recognise the existence of divergent interests and collectivities and which would more realistically attempt to reconcile these interests with the plans of the Board?

Collective bargaining is probably the relationship which allows the constituent representative groups within an industry the greatest independence while at the same time attempting to reconcile their different interests within an agreed set of rules. The weaknesses from which collective bargaining normally suffers are that it is narrow or 'impoverished' in terms of its subject matter, that there are problems in terms of the responsible acceptance of its agreements by those who are to be subject to them and that, particularly in a nationalised industry, there is the converse problem of getting the bargainers to reconcile their members' demands with the requirements of a government concerned with national economic consequences. Other procedural weaknesses have often been identified but they are not very relevant to the situation of the coal industry which is commonly agreed to have, in procedural terms, very effective collective bargaining machinery.

It might be possible to approach the solution of some of these problems, which are amongst the most difficult in the whole field of industrial relations, by attempting to merge the consultative and the collective bargaining process. There have been attempts to extend the influence of joint consultation in recent years, but they do not seem to have been successful. Griffin quotes the 1968 edition of the coal industry guide to consultation, that the committees will be 'required to examine the colliery's business objectives performances' but, he concludes 'this has gone largely unremarked'.[13] Joint consultation purports to be a process of joint control but it is largely concerned with the trivial in which employee interest is low. Collective bargaining is concerned with joint agreement over substantive matters in which employee interest is high but it does not attempt joint control. A merging of the two processes might bring about a situation in which agreements could be made jointly over a wide range of affairs by representative bodies which were responsible for monitoring the agreements which they had made. If workers or other representatives could, in this way, be given a much wider influence in the planning and control of the industry, it might represent a genuine measure of worker involvement and this would be valuable in itself in a nationalised industry and would be nearer to the goal at which joint consultation was ostensibly directed.

Any extension in the collective bargaining relationship will be meaningless unless the process of collective bargaining takes some account of the inevitable influence of the state; if it does not, then collective bargaining begins to lose its significance as the management of a nationalised industry is brushed aside in the preliminaries to the critical negotiations between the government and the union. So we must begin with a discussion of the way in which the state's influence can be allowed for in collective bargaining before we can examine the influence of the worker in a nationalised industry.

This could be achieved by a series of telescoping agreements in which the first stage would be a broad and decisive agreement between the industry's board and the respective government minister. This first agreement would set, for a given period, targets for the industry's achievement in terms of prices, output, markets, profits or deficits and the resources to be made available from public funds for investment in the industry. This overriding agreement between board and government would constrain all subsequent negotiations and trade unionists might therefore complain that it was made without their consent and that it improperly limited their subsequent freedom of manoeuvre. The defence against these complaints would be:

(1) that the state's intervention and overall control of the planning process is necessary and inevitable, that it is best made overtly so that account can be taken of it.

(2) that the state's influence is brought to bear through representatives of an elected government representing both consumers and the union's members.

(3) that the government, in taking its policy decisions in relation to the industry, is open to the quite proper influence which unions or the TUC can exert as pressure groups.

After the first agreement had set the broad commercial goals and constraints within which the industry would operate, the succeeding stages would be no different from the series of contributing business plans by which any large organisation conducts its affairs. The difference is that each of the business plans would be negotiated at the appropriate level between the management and the unions. At national level, an agreement would have to be negotiated, designed to meet the overall requirements arrived at between the government and the industry's board. The national agreement would, in terms of the objectives to be met and the resources available, arrive at output and productivity targets, manpower plans, training requirements, expenditure on health and welfare. Part of the national agreement would determine national wage rates but the rates, like the other constituent parts of the national agreement, would be set within the context of the first agreement between government and board.

Below national level, area and unit agreements would be concluded but each would have to be collectively approved by the higher level negotiating body so as to ensure that the higher agreement would be achieved by the subordinate regions or units which it encompassed.

The existing apparatus of consultative committees could be transformed into standing committees related, by representation, to the negotiating bodies at each level. They could also have a functional relationship in that they would check the performance of each of the parties to the negotiated agreement with powers of inspection which would enable them to monitor progress against plans.

A programme such as this would have several advantages. It would institutionalise and make explicit the relationship of the state and it might go some way to establishing an effective relationship between incomes policy and collective bargaining. It would involve the unions in business planning and give them a responsible share in managerial decisions and in controlling activities directed at their implementation. It would, at the same time, set out to achieve this degree of influence and involvement while allowing the unions to pursue their own interests through the negotiating process without abandoning their essential role and independence. It would allow management and unions to pursue their proper objectives (instead of being persuaded to abandon them in the interests of the other) in a context of greater reality; it is more realistic for union influence on planning to be realised and accounted for before plans, otherwise made in isolation, are made meaningless by isolation. Management might also acquire the advantage of dealing with unions committed to an understanding of business and social constraints and to adopting a more responsible attitude than they are sometimes charged with adopting. Unions might also become more concerned to control and influence the behaviour of their members for the achievement of objectives to the determination of which they have been party.

It may be that such an experiment would be judged to be too radical (although some experimentation in a field in which there has been little or no development for fifty years is surely worthwhile). There are signs of an environment developing which would be more receptive to radical change. While the Board has announced that it will 'open the books' to the Union, the Union has produced a research paper (confidential, but quoted in *The Times*, 20th November 1972) which relates pay bargaining to participation in managerial decision making aimed at achieving joint 'management by objectives which would transfer important new functions to local union officials'.

Whatever pattern is to be developed in the future, it is hoped that we have established that joint consultation has failed to achieve any of the disparate objectives for which it was constructed. As an instrument of management communication to workers it has been replaced by other techniques more completely under management control. As a means of communication and control it does nothing to replace the necessity for sound organisation and for clarifying managerial roles and incorporating subordinate managers (the deputies) within the managerial framework. As a means of achieving unity of purpose, while the goal itself has come to be doubted as a practical objective, there is a growing recognition that informal means of workplace discussion more nearly relate to it. As a means of achieving the ambiguous ends of objective worker participation or industrial democracy, joint consultation must be regarded as totally ineffective. It is at this level, finally, that we have suggested the need for a much more radical or experimental approach, assuming, of course, that there is any serious and abiding interest in the achievement of democracy in work.

REFERENCES

1. H. A. Clegg, *A New Approach to Industrial Democracy*, Blackwell, Oxford, 1960.
2. R. L. Kahn and E. Boulding, *Power and Conflict in Organisation*, Tavistock, London, 1964.
3. A. E. Etzioni, *A Comparative Analysis of Complex Organisations*, Free Press, Glencoe, 1961.
4. National Institute of Industrial Psychology, *Joint Consultation in British Industry*, Staples Press, London, 1952.

5. H. A. Clegg, op. cit.
6. W. D. Rees, 'The Practical Functions of Joint Consultation,' unpublished M.Sc. thesis, University of London, 1962.
7. R. Dahrendorf, *Class and Class Conflict in an Industrial Society*, Routledge & Kegan Paul, London, 1959.
8. W. Paynter, *British Trade Unions and the Problem of Change*, Allen and Unwin, London, 1970.
9. J. H. Goldthorpe, 'Status and Conflict in Industry', Paper delivered to Industrial Sociology Group of the British Sociological Association Conference, 1960.
10. J. Child, *British Management Thought*, Allen and Unwin, London, 1969.
11. Lord Robens, *Ten Year Stint*, Cassell, London, 1972.
12. Alan Fox, *A Sociology of Work in Industry*, Collier-Macmillan, London, 1971.
13. A. R. Griffin, 'Consultation and Conciliation in the Mining Industry', *Industrial Relations Journal*, Vol. 3, 1972.

Employee Directors in the British Steel Corporation

T. KEN JONES*

BACKGROUND

Before considering the more general implications and broader issues raised by the BSC's employee director experiment for the subject of employee participation in the management of organisations, it is necessary to summarise certain of its major features and the developments which have taken place, both directly and indirectly related to the scheme, during the past six years:

1. The Iron and Steel Act 1967 did not require the future Corporation to appoint employee directors and the Organising Committee of the British (then National) Steel Corporation, on its own initiative, proposed the scheme to the Trades Union Congress several months prior to the Vesting Day of 28 July, 1967.

2. The employee director experiment was seen by the Organising Committee as adding an extra dimension to its statutory obligations of developing with the trade unions, effective procedures for collective bargaining and joint consultation. The Corporation and the Unions always recognised that the best way of ensuring the participation of its employees and the trade unions in the management of the industry would be through encouraging trade union membership, extending collective bargaining to groups of employees who had not previously been covered by such procedures, and developing a joint consultative system which was trade union based and very closely linked to the collective bargaining system. Both the letter of the 27 April from Ron Smith, the Corporation's Board member for Personnel and Social Policy, to George Woodcock, the General Secretary of the TUC and the press statement of

---

* The author is an employee of the British Steel Corporation but the opinions expressed are personal and do not represent the official view of the Corporation.

1 May about the employee director experiment, made it clear that the Organising Committee viewed the experiment as an extension to and not in any way a substitute for the traditional methods of involvement.

3. There was no precise definition of the role of the employee directors and the agreed scheme, which eventually emerged as the result of the Organising Committee's proposals to the TUC in April and subsequent joint discussions in June and July, was limited to the following matters:

    a. Up to three employee directors would be appointed to each Group Board.

    b. Appointees would work at their normal jobs when not undertaking their directorial role.

    c. The appointments would be made by the Chairman of the Corporation from a short list of names provided by the TUC from employees of the British Steel Corporation.

    d. Part-time trade union officers could be appointed but they would be expected to resign their trade union positions.

    e. Employee directors would serve on the Group Boards responsible for the works in which they were employed. (The original proposal of the Organising Committee suggested that employee directors should not serve on their own Boards, but the proposal was amended as the result of discussions with the trade unions.)

    f. The employee directors would be appointed for three years and would receive a salary of £1,000 per annum plus compensation for any loss of normal earnings.

    g. The scbeme was to be regarded as experimental and reviewed in consultation with the TUC.

4. The selection process and the first year of the scheme coincided with a period of great strain between the Corporation and the Steel Committee* of the TUC over the white-collar recognition issue in the industry.

---

* The Steel Committee is serviced by the TUC and is composed of representatives of:
Iron and Steel Trades Confederation
National Union of Blastfurnacemen

5. The four Group Boards on which the employee directors sat were formally advisory in nature and the Group Managing Director, who was also a member of the Corporation Board, was responsible for the efficiency of his Group. The Boards consisted of approximately ten full time directors, who either had functional or line executive responsibilities and five or six part-time members of whom three were employee directors.

6. The selection process was completed by March 1968 and employee directors attended their first Board meeting in May/ June. The initial twelve employee directors consisted of:

Union   6 ISTC
          2 NUB
          1 NCC (AEU)
          1 AUBTW
          1 T & G
          1 G & MWU

Union   11 local part time officers, of whom six also sat
Experience on national union committees.

Occupation 5 Production Grades
          2 Craftsmen
          2 Foremen
          1 Senior Clerk
          1 Technician
          1 Middle Manager

7. All the employee directors agreed to attend an initial five-week training course jointly organised by the Steel Industry's Management Training College at Ashorne Hill and the TUC Training College. The three weeks of the course at Ashorne Hill concentrated upon broadening the understanding of the employee directors of the problems facing the BSC and helping them to prepare for their Board Meetings. The two weeks spent at the TUC Training College dealt with such matters as the relationship between Government and industry and an examina-

---

National Craftsmens' (Iron & Steel) Co-ordinating Committee
Amalgamated Union of Building Trades Workers
Municipal and General Workers' Union
Transport and General Workers' Union.

tion of participation in other nationalised industries.

8. Towards the end of 1968, it was decided to initiate a review of the development of the scheme in an attempt to clarify the role and responsibilities of the employee directors on the Boards and then to study their links with the shopfloor and the trade unions. The outcome of the discussions was a job description (Appendix A) which was agreed with the Steel Committee and was to become fully operative when the new product division structure was implemented in 1970. The new structure also provided the opportunity to appoint four additional employee directors. The new allocation was:

	No. of Employees	No. of E.D.s
General Steels	89,800	4
Strip Mills	70,300	4
Special Steels	41,600	3
Tubes	42,100	3
Constructional Eng.	9,500	1
Chemicals	1,500	1

Since it was felt that the new Corporation structure and the job description had created a new situation, it was agreed between the Corporation and the Steel Committee that the experimental period should be extended by one year to April 1972.

9. One of the existing employee directors was appointed by the Minister as a part time member of the Corporation Board in 1970. Although, because of certain statutory difficulties, he could no longer officially be regarded as an employee director, his job description made it clear that he was to have a special relationship with the employee directors and that his main contribution to the Corporation Board was anticipated as being roughly similar to that of employee directors to the Divisional Boards. It was significant that immediately he was appointed by the Minister, the Chairman placed him on the Corporation's Planning Committee and the Finance and Capital Approvals Committee.

10. During early 1969, it was agreed by the employee directors, the Corporation and the Steel Committee that since there

was a commitment to assess the experiment, it would be helpful for an independent review of the scheme to be undertaken. Eventually a two-year research programme, sponsored by BSC and the Social Science Research Council, and supported by the Steel Committee, was agreed. The confidential research report was to be submitted by September 1971. The four members of the Research Team were located at the Universities of Bradford, Sheffield, Strathclyde and Wales (Cardiff), under the general care and surveillance of a Steering Board which consisted of the four researchers, four senior academics from the participating universities, two employee directors and one representative each from the TUC and BSC. The research was also supported by the International Institute for Labour Studies at Geneva.

11. The formal evaluation of the experiment began in mid-1971 and consisted of discussions within the industry, with the Steel Committee, an examination of developments in West Germany and Norway, and very careful consideration of the research report.

12. After searching internal discussions and several meetings between the Corporation and the Steel Committee, the Corporation decided that the scheme should become a permanent feature of its structure. However, the following amendments should be made in order to try to counter identifiable weaknesses:

a. the trade unions could select their nominees by any method they chose, including shopfloor elections, and the Steel Committee's short list should be considered by a joint management/trade union selection committee.

b. the employee directors were to be permitted to hold any lay trade union office.

c. the employee directors were in future to participate actively in trade union/management joint consultative meetings.

d. each employee director would be responsible for a particular 'designated' area within his Division based on geographical and/or product criteria.

Also the job description was rewritten to provide for these changes (Appendix B) and to emphasize the greater degree of involvement the employee directors were to have both in policy formulating committees and with the trade unions.

13. In order to give time to consider carefully the 'designated areas', to allow the selection process to be carried out, and to give time for any new employee directors to complete the training programme (this time two weeks at Ashorne Hill and three weeks at the TUC Training College), it was decided that the new period of office would not commence until 1 April 1973. In the intervening period, the existing employee directors would be once again eligible to stand for trade union office and to attend joint consultative meetings as participants. Six of the sixteen employee directors were to remain in office until 31 March 1975 as they had not been appointed until 1970.

14. The individual trade unions sought nominations during the summer of 1972 and in September the Steel Committee received lists of nominees from the various unions. The Joint Selection Committee which was composed of the Steel Committee and a roughly equal number of Corporation representatives considered 74 nominees in December. All the names forwarded to the Chairman of the Corporation had the unanimous approval of the Committee, but in two cases, the Committee agreed that the Chairman should be asked to select one of two candidates. The names of the ten employee directors were announced in January 1973 and they consisted of five new employee directors and five reappointed individuals.

15. In March 1973, the Corporation announced that the product division boards were to be abolished and employee directors were to become members of the Divisional Management Committees which formerly had only consisted of the full-time divisional directors.

ANALYSIS

Certain features of the BSC scheme highlight a number of issues of broader interest and relevance to the general question of industrial democracy and the more specific matter of employee representation on Company Boards.

Definitions

Perhaps the most fundamental problem which bedevils any rational discussion about employee involvement in management, especially at the Board level, is the lack of any agreed definition of the terminology used. On examination, the individual words: 'employee'; 'involvement'; and 'management', are all capable of a variety of interpretations and taken altogether, the lack of clarity is compounded and the result is confusion and misunderstanding.

Almost everybody thinks he has a clear idea of what an employee is, but the subjective understanding of this word may be rather different from the legal-type definition of 'an individual in employment for which he is paid a wage or salary'. In this sense, the Chief Executive of a nationalised industry or private company, let alone his most senior directors and managers are as much employees as the labourer or junior clerical grades. If this approach is followed, it can be argued that there is complete employee involvement in the management of industry as many private Company Boards consist of a majority of full time salaried executive directors. Consequently a line must be drawn, but where? Certain manual employees would not accept that staff employees require the same form of institutionalised procedures as themselves to ensure their involvement in the management of the enterprise, and many white-collar employees would accept this viewpoint. These attitudes may be disappearing, but the line is only shifting upwards. Should middle or even senior managers in large organisations be regarded as employees in this context?

The next problem is the definition of 'involvement'. The basic principle to which many people would give support is that people who are managed should have some say about the decisions which affect them. Controversy and confusion begin when an attempt is made to specify by what method and to what degree employees should have an influence upon the decisions that affect them.

It is at this stage that discussion usually centres around what are usually characterised as the two basic, but opposed, views of participation or involvement as they relate to the nature of organisations. The Donovan Commission basically accepted the view, so persuasively argued by Alan Fox in his research paper, 'Industrial Sociology and Industrial Relations' that conflict between manager and the managed is both inherent and normal in organisations and that in this situation the only effective way of increasing employee involvement in management is through collective bargaining.

Consequently, the whole question of employee representation on company boards is dismissed in the Report of the Donovan Commission in a way calculated to show that the majority of the Commission regarded the concept as misguided, and the issue as irrelevant. This was particularly unfortunate as it not only tended to stifle intellectual discussion of this matter, but also demonstrated as much as any other part of the Report the limitations of the majority of the Commissioners and the way they had allowed themselves to be bemused by one very able and powerfully reasoned approach to industrial relations.

The approach which is usually regarded as being dramatically opposed is that of those who have tried to analyse organisations from a socio-psychological perspective. The names of this group read like a role of the fads in management theory during the last twenty years—Hertzberg, Maslow, Likert, McGregor. Conflict is seen as unnecessary and irrational and such involvement techniques as mass surveys, job enrichment and enlargement and communications will make all employees realise that the aims, aspirations and interests of management and employees are very closely aligned. It is not without significance that many of the companies who are most closely identified with this approach also discourage trade union membership and have resisted the recognition of trade unions for collective bargaining purposes.

In the confused world of reality, it is not impossible for business organisations and trade unions to espouse both sets of views consecutively over short periods of time or even simultaneously! The attitude of the TUC has changed significantly with regard to the problem of employee participation at board level during the last ten years. Similarly, the CBI might also be considered to have moved from a paternalistic 'all in the same boat together' approach to one which recognises that there are legitimate differences of views which may result in the Captain and Officers wishing to go one way, but the rest of the crew holding rather contrary views. Consequently, it is not surprising that it is possible for an employer to hold a pluralistic conflict view of industrial relations whilst supporting the development of ideas advocated by those with a unitary view of organisations.

The final problem of definition is what is meant by 'management'. The obvious question is 'management' at what level—the shop floor, the board room, at some intermediate stage, or all levels? There are a large number of studies both in this country and abroad, which have been able to show that the majority of non-managerial employees are primarily interested in being involved in management decisions which have a reasonably immediate effect upon their working lives such as changes in the method of payment, disciplinary procedures and working hours. Also, not unexpectedly, only a comparatively small number of employees are interested in such problems as the allocation of profits between investments, dividends, or reserves, the long term development planning of the organisation, or commercial policy. Such conclusions by themselves tend to demonstrate the predilection of social scientists to rediscovering the wheel for the nth time. The value of answers to hypothetical questions about situations which are completely beyond the experience of the respondents is dubious in the extreme and the data can sometimes be used to provide arguments to support the prejudices of the researcher.

As pointed out earlier the Corporation publicly declared its attitude to the question of what were the ways its employees were to be involved in the management of the Corporation. Also an attempt was made to establish up to what level in the organisation these institutional procedures were to be applied. The answer was up to and including Assistant Departmental Manager, thereby excluding some 2,000 to 3,000 of the 260,000 employees from participating in collective bargaining, joint consultation, or the employee-director scheme. However, in an industry with such enormous variations both in size of production units, technology and traditions, this definition did not represent a fine line, but a smudged broad grey area over which were to be fought some very bitter recognition battles.

The Introduction of an Employee Director Scheme

The basic question in its simplest form is: should it grow from the bottom upwards or be introduced from the top?

An approach of the latter sort clearly has certain major disadvantages compared with the former organic growth concept. It can be interpreted as a clever managerial device to prevent a more radical demand winning support and consequently it may be opposed by some trade union leaders and by local activists who may have their own ideas as to how such a concept should be implemented. Furthermore, the communications problem will be far greater unless there is a broad desire for, and an expectancy of, such a development. But in spite of these disadvantages, an initiative from the top of the managerial hierarchy in response to a general pressure from the trade unions and the largely unformulated desire of the employees for a development of this type, has two advantages.

First, it has a far better chance of being introduced reasonably quickly, and second, since the development in all probability will have the support of the Chief Executive or Chairman of the organisation, it stands a better chance of not being frustrated by management. The response of many trade union leaders to such a top level management initiative is likely to be that of the Steel Committee of the TUC: view it with caution, influence management to amend certain features of the scheme, adopt a 'wait and see' approach. This is understandable but exacerbates the difficulties employee directors may have in their relations with the trade unions.

A far better method of introduction is likely to be a joint problem-solving approach. Such an approach does not commend itself to the ideologies on the trade union or management side, but it is possible for such an approach to be used in a way which is compatible with a conflict view of industrial relations. The subject for examination should be the whole question of participation and out of this one of the recommendations may well be that consideration should be given to introducing an employee director scheme. The detailed provisions of the scheme should be jointly agreed between management and the unions. Such a method is more likely to produce a scheme which is closely related to the nature and industrial relations character of the industry and the predilections of the management and trade unions concerned.

Should the British Parliament pass legislation to make a two-tier Supervisory and Executive Board structure obligatory for all companies above a certain size with employee directors on the Supervisory Board, a different and more difficult situation is created.

A Gallup Survey in 1969 found a large majority of employees in favour of participating at Board level and surveys conducted both in British Rail and British Steel showed about 70 per cent of employees in favour of the concept of employee directors. However, general employee approval could soon give way to disenchantment if a statutory scheme was introduced to which many trade union leaders had an ambivalent attitude and which many managements felt they should attempt to circumvent.

Selection Process to Supervisory Boards

Any scheme for employee directors must provide for a selection process and the most obvious way of undertaking this is by direct election from amongst the employees. In theory, this seems easy, but in practice it may well provide almost insurmountable problems.

In a one-plant company which only recognises one union for collective bargaining and consultative purposes, the only problems are concerned with occupational representation and the question of the representation of non-union employees. On the assumption that the Industrial Relations Act does not present legal obstacles, then in a situation in which trade unions have recognition and preferably agency shop agreements for all occupational groups, only trade unionists should participate in the selection. As soon as more than one union is involved, or a major occupational group is not unionised, it is likely to be necessary to develop some form of allocated representation.

In a multi-plant company it will be very difficult to stop the large plants dominating the election process and when the company has plants in a large number of different areas, the question of regional representation may be raised. If all these factors are to be taken into account, the result may well be an electoral system of bewildering complexity and administrative unwieldiness.

Therefore, if there is to be an electoral system for employee directors in a situation in which trade unions are recognised for all occupational groups, there is a strong case for arguing that it should be on a multi-union constituency basis and it would be desirable for the elections to be held by postal ballot and conducted by an independent body. Special arrangements would have to be made for any major non-unionised occupational group. The definition of the constituencies would be jointly agreed between management and the trade unions. In non-unionised organisations, an independent body would have to establish the constituencies and make the necessary arrangements for the electoral process.

This approach would be compatible with the findings of the employee director research team, that a large number of the trade union activists saw the employee directors as being over and above individual trade union loyalties and very few people suggested that appointments should be made by one union only. This view would certainly be endorsed by the BSC and employee directors, who were very concerned not to appear biased in favour of the interests of their own union. Both versions of their job description provided for them to attend meetings of trade unions other than their own. In practice, certain of them eventually succeeded in being regularly invited to meetings, at a variety of levels, of production, craft and general unions. In an industry in which, prior to nationalisation, the full time officials of the major unions had never sat down together to discuss any type of common policy, such a development can be considered a breakthrough.

Although such an electoral system might seem to commend itself wholeheartedly to the trade unions, it does pose certain problems. An employee director who is also an influential lay official, chosen by this direct election method, might well appear to the ordinary employees, far more influential than the full-time officials, not only upon day-to-day matters of his plant, but also upon determining the general shape of the company's policy and the way it is implemented at plant level. Furthermore, the extent of multi-union support which such an individual might have, although familiar in some industries, would be foreign to many others. A further potential complication is that a direct electoral system would firmly establish *the* representative link and it might make it, not impossible, but more difficult to keep the responsibilities of the trade unions for collective bargaining separate from the responsibilities of the employee director. The present system in the BSC provides, if the unions so desire, for the ordinary membership to elect the nominees and for the national trade unions to be party to the selection process, but in spite of their continuing to hold union office, there can be no suggestion that the views of the employee directors, as presented in the Boardroom, are anything more than *a* representative view of shop floor opinion. *The* representative view of the trade union members is expressed by the full- and part-time officials.

For those who implicitly believe that the collective bargaining approach should be carried into the supervisory board, the only acceptable system is likely to be the direct multi-union constituency election of employee representatives to fill 50 per cent of the seats. However, unless it is made clear that the terms of reference of the supervisory board exclude such matters as pay and conditions of employment, the implications for the national trade union officials and the organisation of their unions may be far more widespread and more unpleasant than they anticipate. The election process combined with joint selection would enable a representative voice of the employees and, informally, an official trade union viewpoint, to be influential without transposing the legitimate union-management conflict situation to the supervisory board or possibly imposing constraints upon future union action through the collective bargaining system.

Role Conflict

The feature of employee director schemes which causes considerable anxiety amongst management and trade unions, and gasps of perplexed horror from many sociologists, is that of role conflict. This was certainly considered by the Organising Committee of the BSC in early 1967 and it was to blunt the sharpness of this potential conflict that it was proposed that employee directors should be appointed to boards not responsible for the works in which they were employed and asking employee directors to resign their trade union offices.

This latter condition certainly helped to modify the type of role conflict which may have arisen from holding an influential position in two institutions which might be involved in destructive as opposed to constructive conflict. However, this condition did not necessarily lessen the type of personal conflict which could result from an individual being on both sides of an issue at the same time.

The research team was able to show in a reasonably conclusive way that as time went by, many of the employee directors began to be influenced more by the management viewpoint than by the trade union approach to particular issues. This was largely because their links with the trade union activists had been severed whereas their connections with management were gradually strengthened. Consequently, it could be argued that the degree of feared role conflict failed to emerge, not because of the skill and adaptability of the individual employee director, but because the employee directors ceased to fulfil a role as representatives of the workforce and merely became part-time directors who happened to work for the BSC.

Within a few years of taking up their appointments the majority of the employee directors, whilst still unquestionably concerned with representing the interests of the employees, regarded their prime responsibility as contributing to the efficiency of the industry, but this was not surprising since the research team also found a basic commitment amongst the ordinary employees and the lay trade unionists to making the steel industry viable. The employee directors held a unitary concept of the BSC but in this they reflected the views of the majority of the employees and the trade union activists.

Therefore, in the steel industry role conflict did not come to a head over general issues but this may not be the case in other industries where a large percentage of employees may have a less well developed 'stake in their job'. However, there were many instances of pointed personal conflict, in particular, in connection with union recognition disputes, the rationalisation programme and industrial disputes in which employee directors were personally involved. In these situations their experience of sometimes facing roughly similar situations as trade union officers when their local members may have been in conflict with the official union policy and the common sense of ordinary employees, full- and part-time officials, management as well as the employee directors, enabled them to cope with the situation. Most of the employee directors soon learnt the truth of the statement made by that leading Clydeside rebel of the 1920s, Jimmy Maxton: 'If you cannot ride two bloody horses, you don't deserve to be in the bloody circus.'

Evaluation and Changes

It is doubtful if any internal evaluation of an organisational change can be sufficiently objective, especially if the individuals responsible for approving any changes are basically the same people who established the initial scheme. In this situation it is particularly valuable to enlist the services of outsiders who can be impartial within the constraints of their own prejudices.

The research was conducted, as anticipated, with great professional expertise but of even more importance, it was done with tact and understanding of the danger that their very presence might influence the behaviour, not only of the employee directors, but also the other people who came into contact with them. This was a particularly difficult problem in connection with the observation of Board meetings and it is remotely possible that the presence of researchers resulted in certain issues being discussed at Management Committees and not at the Boards.

The research report, in spite of the almost inevitable hold-ups, in the computer analysis of the data, was received on the day required. It was separately discussed with representatives of the Corporation and the Steel Committee. Although the Corporation felt that there were certain weaknesses in the report, the basic criticisms that it made in connection with the appointments system and the general relationships with the trade unions were accepted.

In the meetings held between the Corporation and the Steel Committee it was accepted that the research team had correctly diagnosed the major weaknesses in the structure of the scheme. Although consideration was given to the possibility of direct elections from the shop floor to the Boards or placing responsibility for selection entirely in the hands of the unions, it was soon

resolved that the best system would be that described earlier
(page 87). This method undoubtedly provided for a greater
degree of participation by the ordinary employees than hitherto
and highlighted the fact that the selection was a joint union-
management process. It is a matter of some regret that none of
the unions concerned decided to use the direct electoral system
to provide nominees.

Once the issue of the selection process was resolved, attention
was concentrated upon the question of the holding of trade
union office. The initial Corporation viewpoint was that employee directors could hold any office other than that of the
leading negotiator at plant level (for example, a craft employee
director could be a shop steward but not a works convener),
or hold an executive council or national negotiating position.
The unions were divided and some certainly saw that it would
be extremely difficult for an individual to be able to give the
service required of a senior part time officer to his members,
whilst fulfilling his obligations as employee director, which
might result in his being away from the plant for considerable
periods. Eventually, it was agreed that an employee director
could hold any trade union position and that this issue should
be discussed thoroughly between the employee director desig-
nate and his members.

The amendments made to the scheme should improve its
effectiveness but as in any organisational change, the improvements will undoubtedly create new problems. The first problem
is that the new employee directors are now faced with a difficult
personal decision whether or not to resign any of the trade union
offices. One has resigned from the national negotiating committee of his union, one is staying on his national executive committee, but another has resigned a similar position. These are
temporary difficulties but of more fundamental importance is
that these employee directors may well be faced with exceed-
ingly difficult role conflict situations—they will not just need to
be skilled equestrians but contortionists as well.

Successes and failures

One of the consistent objectives of the employee director
scheme has been to bring to the Group Divisional Boardroom *a*

shop floor viewpoint. At no time was it suggested that the employee director would represent *the* shop floor viewpoint, this was the responsibility of the trade unions through extended negotiating and joint consultative machinery. When the scheme started the employee directors were able to present *a* shop floor viewpoint, not only because at a personal level they continued to work at their normal jobs for at least an average of 20 hours per week, and since all but one of them had held trade union representative positions, they could speak with some authority about the views of the active trade unionists.

Although continuing to undertake their normal work has sometimes created practical difficulties for departmental management and their work contacts are sometimes very limited, it probably has made a considerable contribution to preventing them as individuals becoming remote from shop floor experience. This system has avoided the situation sometimes found in West Germany whereby certain employee supervisory board members who are also works councillors may not have had practical work floor experience for many years. This is not to argue that in order to represent employees an individual must work on the shop floor; the official representative view of the opinions of trade union members can be put by full time, as well as part time, trade union officials; but if one of their jobs is to represent a shop floor viewpoint, there is some sense in having employee directors who still have a practical feel of what it is like to work on the shop or office floor.

Because the employee directors had to resign their trade union positions, it meant that they gradually ceased to have a detailed knowledge, arising out of their own activities, of the views of the trade unions and this resulted in their opinions becoming progressively more personal as opposed to representative.

During the experimental period various attempts were made to correct this situation and some of the employee directors were remarkably successful in winning the acceptance of part-time union officials to the extent of their being invited to attend a far broader selection of union meetings than they ever did prior to becoming employee directors. However, as pointed out earlier most of the employee directors came into contact with management views more than those of trade union officials and this probably resulted in them becoming less representative than they had been originally of the views of the ordinary employee. Since the major contribution of the employee directors to the Boardroom discussions was that they represented a viewpoint based upon different experiences and values from that of the professional managers, this was a weakness in the scheme.

The other major, publicly declared, objective of the scheme was that it could result in a greater involvement of employees in working out the future policies of the Corporation. However, policies in the BSC are not decided by Group/Divisional Boards —this is the prerogative of the Corporation Board on which there is no official employee director, but one of the part-time Board members is an ex-employee director. The job of the Divisional Board is to advise the Divisional Managing Director, who can influence the development of Corporation-wide policies and who also has considerable latitude over the interpretation of Corporation policies within his division. Furthermore, it is at the Divisional level that many of the major development proposals are initiated.

Whilst the activities of the employee directors were limited to attending board meetings, they could have comparatively little influence, but gradually they began to become more involved in sub-committees and working parties, who evolved policy both at national and divisional level. It was unfortunate that this type of activity was only spasmodic at the time when the research team was gathering its evidence, but employee directors are now to be found, for example, on Divisional Planning Committees, Personnel Advisory Committees, Commercial Assessment Committees, Social & Regional Policy Committees and they have participated in a large number of working parties of which two have been chaired by employee directors.

The research team also raised the question as to whether the Divisional Board was the right forum for the employee directors, especially since the full-time directors also met separately as a Divisional Management Committee. The Divisional Boards have now ceased to exist and the employee directors are members of the Divisional Management Committees.

Consequently whilst it was true that during the early years of the experiment the involvement of the employee directors in the formulation of policies was severely limited, there has been, especially during the past two years, a development which has resulted in employee directors being more involved in an ever broadening range of committees where policy is evolved. They are also members of the committee which the research team regarded as the most influential group in the divisional structure.

Ordinary employees may find it difficult to appreciate the significance of such developments but active trade unionists will well understand the importance of this change. However effective the collective bargaining and consultative machinery, trade unions are usually placed in the position of entering into the discussion at a time when the ideas of management are already well formulated. It may be possible to change these ideas in the consultative and negotiating process but the time when they can be easiest influenced is when they are still very fluid and before they are endorsed by the appropriate management committee. This is the stage where employee directors can regularly be involved but trade union officials cannot— apart from the type of occasional joint working party exercise which certain companies and unions developed to work out productivity agreements. Therefore, the employee director experiment has, possibly in a limited but significant way, advanced the boundary of employee participation into areas which primarily had remained within the prerogatives of management.

Judged by some of the public statements made by the Organising Committee in 1967, about the objectives of the employee director experiment, it can be concluded to have failed in several ways. Judged by the progress which has been made in developing and improving the original rather vague aspirations and establishing a firm basis for future development, it may be regarded as being a success. Judged by the amount of interest it has stimulated amongst those in management, trade unions and government, both in the UK and abroad, and the problems and possible solutions it has drawn attention to, it must be evaluated as an interesting experiment which has made a valuable contribution to the discussion of industrial democracy.

CONCLUSIONS

The employee director experiment was a modest attempt to introduce a new level to the institutionalised procedures for involving employees in the management of industry. Although the National Craftsmens' Co-ordinating Committee for the Iron and Steel Industry had advocated a more radical scheme of this sort, and a number of iron and steel trade union study groups in the Sheffield and Scunthorpe areas had developed a carefully agreed programme for workers' control in the industry, there was no perceivable general demand for such a development from the unions which represented the majority of the employees or from the shop floor. Needless to say there was very little enthusiasm for the idea from the directors and managers of the old private companies, who at best saw the scheme as a rather eccentric development and at worst considered it to be an attempt to modify the traditional power structure within the industry.

The experiment was introduced at a time of great turmoil within the management structure of the industry and in a period of considerable bitterness between the traditional steel unions and the Corporation. Furthermore, this was a time when the Head Office had only rudimentary central control over the internal communication media. Until the reorganisation of the Corporation took place in 1970, the role of the employee directors was rather ill-defined and certain of the leading figures in the old Company Boards were, whilst not openly hostile, apparently fairly determined not to involve the employee directors significantly in policy formulation. During the same time their relations with the trade unions at national level were usually very distant and at local level there was often suspicion of the employee directors.

The situation changed significantly for the better in 1970. At first the improvements were slow but major changes began to be apparent from about March 1971. The employee directors' relationship with the trade unions both at national and local level improved considerably and at board level they began to become involved far more in influencing the development and implementation of policy. However, there were still serious

weaknesses within the scheme but these were candidly recognised during the evaluation process and important amendments made to the scheme. The second stage of the scheme
which started in April 1973 should mark a further advance in
its development, but it will also pose certain additional critical
problems which will have considerable implications for the
development of employee directors on supervisory boards in the
private sector.

The success achieved by the experiment was due largely to
the support it received from the Chairman of the Corporation,
Lord Melchett; Mr. Ron Smith, the Board Member for Personnel and Social Policy; a few of the national trade union leaders,
in particular Mr. John Boyd, AEUW; the encouragement and
goodwill of many Divisional Directors and Managers and local
full- and part-time trade union officials. Without this support
the scheme would probably have foundered whatever its merits.
However, an even more critical factor in contributing towards
such success as the scheme enjoyed was the ability and common
sense of the employee directors. They were placed in situations
which demanded of them the greatest tact possible, together
with a resolution to ensure that the scheme was gradually improved. Although mistakes were occasionally made, their relationships with management, the trade unions and the employees demonstrated their integrity and ability.

An organisational change of this character is a painful and
difficult process and it takes many years to develop its potential.
For a long time progress may seem to be almost negligible and
after five years of the BSC's employee director scheme it was
only possible to point to a few unquantifiable benefits derived
by the employees and management and at the same time it was
also possible to highlight certain weaknesses. However, the real
success of the experiment was that it survived and established
a basis for further developments in employee participation in
the BSC, not only in the employee director area, but also over
a far broader spectrum which included a variety of techniques
to introduce greater participation at shop floor level to comple-
ment the collective bargaining and joint consultative procedures, and the involvement of employees at the level of the top
management committees.

Appendix A December 1969

Basic Function

As a member of a Divisional Board to assist the Managing Director in his responsibilities for directing the affairs of the Division, particularly by seeking to present the point of view of his fellow employees thus contributing to the involvement of the employees in the management of the British Steel Corporation.

Main Responsibilities

1. To perform his normal job as an employee.
2. To fulfil his role as a non-executive director by:
   (a) attending and participating in Divisional Board Meetings.
   (b) being available for consultation with the Divisional Managing Director.

   and also by

3. Attending and participating in such of the following activities as may be decided, after consultation with the employee-director, by the Divisional Managing Director:
   (i) Standing Advisory Committees, working parties and study groups at Divisional level.
   (ii) Formal and informal meetings of functional directors and local management.
   (iii) Special studies or duties at home and abroad.

4. To prepare himself for effective participation as an employee-director.

5. To draw on his experience and the views of fellow employees when acting as an employee-director by:

(a) Attending his own trade union's meetings within the Division.

(b) Attending, by invitation, other trade union meetings as observer within his Division.

(c) Attending, as observer, joint consultative meetings within his Division.

(d) Visiting establishments in his Division and meeting employees and management.

6. To prepare a monthly programme of his activities for submission, through his local management, to the Divisional Managing Director for his endorsement. In the event of having to undertake unscheduled duties as much notice as possible should be given to the Divisional Managing Director through the local management.

Appendix B January 1972

BRITISH STEEL CORPORATION
JOB DESCRIPTION—EMPLOYEE-DIRECTOR

Basic Function

As a member of a Divisional Board to assist the Managing Director in his responsibilities for directing the affairs of the Division, particularly by keeping in close contact with the joint consultative committees and the trade unions in the BSC and contributing to the involvement of the employees in the management of the British Steel Corporation.

Main Responsibilities

1. To fulfil his role as a non-executive director by:
   (a) attending and participating in Divisional Board Meetings.
   (b) being available for consultation with the Divisional Managing Director.

   and also by

2. Attending and participating in such of the following activities as may be decided after consultation with the employee director by the Divisional Managing Director:

(a) Standing Advisory Committees, Working Parties and Study Groups at Divisional and lower level.

(b) Formal and informal meetings with other divisional directors and local management.

(c) Special studies and duties at home and abroad.

3. Participating in such national working parties and advisory committees as may be agreed between the relevant Board Member or functional Managing Director, the Divisional Managing Director, and employee director.

4. To prepare himself for effective participation as an employee director.

5. To draw on his experience and the views of fellow employees, in particular in his designated area, when acting as an employee director by:

  (a) Continuing to hold appropriate part-time trade union office, when requested by the local membership of his trade union.
  (b) Attending meetings of his own trade union.
  (c) Attending meetings of other trade unions.
  (d) Attending and participating in joint consultative meetings.
  (e) Visiting establishments and meeting employees, trade union representatives, and management.

6. To keep in touch with trade union views and maintaining an understanding of trade union policies in the steel industry by:

  (a) keeping in contact with the full-time officials of the appropriate trade unions.
  (b) attending meetings when requested of the Executive Council, or such bodies as national conferences of trade unions, in order to discuss the work of employee directors.
  (c) participating with all the other employee directors in discussions with the Steel Committee when requested.

7. To perform his normal job as an employee when not engaged in any of the above activities.

8. To inform local management of his programme for the succeeding month and to give as much notice as possible of any unscheduled engagements. To discuss, in advance, with his Divisional Managing Director, participation in any extraneous activities which could affect the Corporation.

Shipbuilding

K. J. W. ALEXANDER

In the United Kingdom until recently workers in the ship-building industry exhibited very little interest in industrial democracy and the workers' control movement. The literature of industrial democracy deals hardly at all with shipbuilding, and the speeches at and delegates to the large conferences of the Institute of Workers Control did not come from shipbuilding workers. Quietude on this issue—which was attracting increasing attention from workers and unions in many other industries in the 1960s—had not always been a feature of shipbuilding. In one of the very few writings specifically on industrial democracy in shipbuilding (though coupled with engineering) G. D. H. Cole had singled the industry out, along with engineering, the mines, the railways[1] and the Post Office, as one in which the demand for control was most vigorous. The vigour was a by-product of the growth of the shop-steward movement in shipbuilding during the First World War, and the interest in control fairly quickly found its way back into the more limited objectives of the unions.

Two recent developments on the Clyde have related the shipbuilding industry to the issue of industrial democracy and workers' control—the Fairfields experiment[2] and the Upper Clyde work-in. This chapter will deal with each of these, as well as drawing upon the experience of self-management in a Yugo-slav shipyard. A final section will speculate upon the forms which industrial democracy could take in the shipbuilding industry in Britain.

INDUSTRIAL CHARACTERISTICS AFFECTING ATTITUDES TO CONTROL

The work processes of shipbuilding make tight managerial control of work-pace and organisation extremely difficult. Incentive schemes and finishing bonuses have been developed as substitutes for management control. The scale and diversity of the assembly process which is what modern shipbuilding largely is, the inevitable remoteness and, indeed, privacy of many of the work-tasks from each other and from continuous or frequent managerial supervision, have contributed to a certain freedom of action coupled with independence of spirit of many ship-building workers. Innumerable work practices have developed and become firmly established in response to the preferences of individual workers and small working groups. When manage-ments have been aware of such practices and disapproved they have often found it impossible to bring about a change, and very frequently no attempt was made. This high degree of influence over their work-performance which many shipbuilding workers can exercise has not, however, evolved into any more formal movement for workers' (or trade union) control over or par-ticipation in managerial decision-taking. The notion of 'en-croaching control', extending from simple decisions about the pace or organisation of work of the deployment of the existing labour force to the more complex and fundamental aspects of management, such as employment and investment policy, has not been borne out in the experience of British shipbuilding. The reasons for this lie partly in the nature of the control already encroached upon, partly in the economic circumstances of the industry and partly in the structure of trade unionism in ship-building.

The control already established lies with individual workers and small groups and squads of workers, and it differs very con-siderably in character and significance between jobs, and for different trades and grades of labour. This absence of uniformity in the forms of control by workers over their work is a feature of many industrial processes, but is probably extreme in the case of shipbuilding. Certainly it has been one of the factors pre-venting any generalised view of the importance of such control emerging amongst shipbuilding workers, and that in turn has held back any demand to institutionalise and formalise the character of control in the industry. Also holding back the formulation of such radical demands has been the very special nature of the demand for labour and of trade union bargaining

power in shipbuilding, which has been derived from the cycle
of production involved in shipbuilding and the cyclical character of the demand for ships as a whole. The production cycle in
shipbuilding, running from design and drawing-office work,
through materials requisition to steel cutting and assembly (some
in the shops, some on the berth and some at the quayside) and
to outfitting and finishing creates sizeable fluctuations in the
demand for different trades and grades of labour at different
times, and this has been one factor producing the acute section-
alism of organisation and attitude amongst shipyard workers.
Redundancies arising from the phases of this production cycle
have been accepted, unlike threatened major shut-downs, as
natural phenomena, and in affecting different groups at different
times, their incidence has tended to separate and distinguish
the interests of these different groups rather than unite them.
The uneven incidence of the opportunity of very high earnings
from finishing bonuses designed to induce maximum effort to
meet a launching or trials deadline has also been a divisive
factor arising more from the physical characteristics and commercial hazards of the industry than from any attempt by
managements to 'divide and rule'. The cycle in the demand for
ships introduces a further factor inhibiting the development of
a common policy amongst shipyard workers. At times of low
demand both unemployment amongst shipyard workers and
keen competition between shipbuilders for orders has tended to
reduce the bargaining power of shipyard unions. These are the
circumstances which it is now widely recognised have bred the
emphasis on narrowly defined job protection amongst tradesmen, a protectionism aimed more at other trades and grades of
labour who could threaten jobs by performing some tasks incidental to their own but the preserve of others, than it is aimed
at shipyard management. Some part, that is, of the control
exercised by shipyard workers over their work is aimed at
safeguarding their jobs against encroachment from other
workers, and this pronounced characteristic of workers' control
in shipyards has been a major factor preventing the extension
and formalisation of workers' control in the industry. It is
significant, in fact, that the elements of workers' control in
shipbuilding which have been formalised and brought to the

attention of managements by union representatives are those of this protective and sectionalist kind. This segmented character of trades unionism in shipbuilding, more so in yard bargaining than in national bargaining, reinforced by the unevenness of the existing incidence of control by workers over their work and of the shifts and changes in the demand for different sectors of the labour force, together explain the failure to advance—even at the level of policy demands—beyond the point of individualistic and small-group control over work situations in shipbuilding.

THE FAIRFIELDS EXPERIMENT

In October 1965 the Fairfields shipyard in the Clyde collapsed financially. The shop stewards conducted a powerful campaign to convince the Labour government to provide finance to keep the yard open, and were helped in this by the offer of a local industrialist—Sir Ian Stewart—to run the yard as a national proving ground. The aims of the proving ground idea were to demonstrate that industrial relations in shipbuilding could be improved by high and stable earnings which were to be made possible by a combination of modern management techniques, effective communication, better utilisation of labour, manpower planning, the extensive use of retraining, the abandonment of restrictive practices and the voluntary relinquishment of the strike weapon. There was a hope that if success of this sort could be demonstrated in what was recognised as being a particularly difficult industrial situation, it could be transmitted more widely in British industry. It was described by George Brown, First Secretary of State at the Department of Economic Affairs as:

> a quite new partnership not only between Government, private enterprise and the two unions, the motive being not merely to save a recently modernised Scottish shipyard from extinction—but in addition to provide a proving ground for new relations in the shipbuilding industry which could change the whole image of our country.[3]

The original emphasis upon an extension of industrial democracy was small. A mass meeting of the workers in the yard had the proposals put to them for a new style of management and for improved methods of work and increased productivity. This meeting endorsed a no-strike pledge as part of the policy for the yard's future. Investment by five unions gave the notion of 'partnership' an initial boost, and this was extended by two other factors. The first was the nomination to the Board by the investing unions of a trade union leader to represent their interests; in addition another trade union leader served on the Board as a nominee of some of the private investors. The second was the limited commitment of the new top management recruited by Stewart to extending workers' participation, which took many forms and the pressures which led to an extension of this commitment, although commentators and observers outside the yard frequently interpreted the experiment as a major shift toward workers' participation and even control. This was neither the intention nor the reality within the yard. The most compelling factor extending worker participation was the wide-ranging changes which management sought to introduce in a short period of time and the requirement, therefore, of management to explain and consult on these changes. Against the background of the establishment of the new company, owing a great deal to shop-steward campaigning and trade union investment, the natural tendency for such communication and consultation to merge with the decision-taking process was accentuated and to some extent formalised. Thus the extension of industrial democracy within Fairfields was primarily a response to management's need to expedite change rather than a response to the shipyard workers desire to participate or control, and confirms rather than confounds the apathy about formalised control which it has been suggested above workers in the industry exhibit.

This conclusion should not be interpreted as meaning that there was a lack of response to opportunities to participate which evolved within the changing managerial situation in the yard. The response was enthusiastic, positive and constructive. The lack was in initiative to introduce or extend new forms of worker participation.

The extension of participation by shipyard workers in the

running of the yard is best outlined by describing the framework available for such participation.

The most remarkable exercise in communication was the
five mass meetings which heard reports from the Chairman and
gave rank and file workers as well as shop stewards a chance to
ask questions and put points of view on the company's position
and policy. It is obvious that meetings attended by over 2,000
workers could not be vehicles for decision-taking, but they
reflected a determination by management to 'take the pulse'
and speak directly to the individual, and also symbolised the
extension of participation which was taking place in more
practical forms at other levels.

Secondly a network of briefing groups was established which
enabled major policy decisions to be conveyed to all levels of
employee within a few hours of the decision having been taken.
Although the main flow in this chain of communication was
from the top downwards, there was a reverse flow or feed-back
which itself reflected a higher level of involvement than had
existed before.

The third manifestation of participation by workers in
management was the greatly enhanced role of shop stewards and
trade union representatives generally within the yard. There
was a full-time convener and a full-time steward representing
the boilermakers' interests, with office accommodation and
practically unlimited facilities and opportunities to consult with
their fellow stewards and members and to have access to
management. Most importantly the convener and two other
stewards attended meetings of the Executive Management
Committee, heard the reports from the Managing Director and
other senior managers to that Committee and participated in
the policy discussions which followed. In addition shop stewards
were heavily involved in committees carrying through a job
evaluation exercise. This participation in the Executive
Management Committee gave the senior stewards full insight
into the current problems of the yard and the alternative ways
of dealing with problems which were under consideration.
Their presence enabled management to gauge the likely
response to proposals and undoubtedly led to better decisions.
It may be argued—and indeed the Managing Director at

Fairfields would himself have argued—that the line between consultation and direct decision-taking had not been breached, and that the right to take decisions still lay with top management. In theory this was so, but this view rests upon a view of decision-taking as a singular event rather than an end-product of a process in which the search is for consensus—or a workable solution. The new relationships and structure at Fairfields had considerably increased the impact of employee thinking and objectives on this process of decision-taking, and both the desire and the necessity of finding consensus solutions was also increased by the circumstances of the yard. The circumstances created a momentum for increased democratic participation which gained rather than waned throughout almost all of the short two years of the Company's independent existence. As has already been indicated this gain in momentum could have been accelerated had the trade union representatives strongly desired this and pressed for it. But although the response was positive it was not followed through by such pressures for increased participation.

The remaining two elements in the framework of increased participation by employees and unions in the affairs of the Company were at levels above that of shipyard production, although both directly concerned with it. First there was the Central Joint Council, on which the main unions were represented along with four management representatives and from this Council four trade union members and two management representatives formed the Scottish Conference to which all unresolved matters had to be referred. In eight of the nine cases referred to this Conference the claim coming from a trade union (or group of unions) was rejected and the management view upheld. This record is powerful evidence against the view that trade union participation in decision-taking is bound to lead to sharper conflicts or too liberal settlements, and supports the view that responsible actions flow (and can only flow) from situations in which individuals and institutions are given an opportunity to exercise responsibility.

The second element is the framework of increased participation which apart from the yard and its network of points at which participation could take place was the Board, and the membership of two trade union leaders on it. This had little impact on yard affairs. Indeed it was remarkable that no attempt was made by the shop stewards to lobby or otherwise pressurise these trade union directors. The presence of trade union leaders on the Board played some part in reassuring the shop stewards that top-level decisions could not be sharply at variance with the interests of the workers; it was clear that the reputation of the trade union leaders was involved. It also provided top management with additional information on trade union attitudes and concerns, which played a part in guiding decisions in a direction more likely to receive support.

This total participation package at Fairfields was more a response to the need to stimulate change and arrive at agreement on how best to bring it about than a conscious attempt to construct a new structure relating employees and their representatives to managerial decision-taking. It was probably more effective because it was designed in a somewhat *ad hoc* way and in response to perceived need rather than received doctrine, but inevitably it would have had to be more adequately integrated had it been allowed to develop over a longer period than in fact was possible due to the incorporation of Fairfields in UCS in 1968. Already there were tensions developing between the shop stewards and the day-to-day participative process they were involved in and the full-time trade union officials' representation on the Central Joint Council. Links were being developed between these two levels which in time would probably have been formalised and led to a clear division of labour between the two, and the possibility of the performance of some overlapping functions, involving representatives from each level. It seems at least possible that the evocation of the trade union directors from all other levels of participation would also, in time, have come under pressure and been qualified to some extent—perhaps by the creation of some reporting and questioning channel which would have been operated between trade union representatives involved at the three levels of Board, top executive management and yard management. An extensive attitude survey conducted at the yard demonstrated that the impact which the creation of these new channels for participation had had on rank and file workers was very considerable. It is difficult to disentangle the various influences of a new management, the new role for trade union representatives, the introduction of a new system of incentives and the trauma of the hairsbreadth escape from final shut-down, but there were marked improvements first in morale and later in productivity and the impression of close observers was that the participation of trade union representatives in management decision-taking played a considerable part in those improvements. There is of course no way of proving or disproving such an impression, but it can at least be suggested as likely that without that one change in the degree of participation these improvements would not have been achieved.

THE UCS WORK-IN

In June 1971 Upper Clyde Shipbuilders Ltd. applied to Government (a major shareholder with the Shipbuilding Industry Board) for £6m. aid to stay in business. When the aid was refused the Company went into liquidation on 14 June. At once the UCS Shop Stewards Co-ordinating Committee initiated a campaign to save the four yards (at Govan, Fairfields, Linthouse, Scotstoun and Clydebank). This campaign was along similar lines to that successfully conducted by the Fairfields stewards for the Govan Yard at the close of 1965, but its tempo was more intense and its scale greater, with a half-day strike of 100,000 supporting workers on June 24. This shift in emphasis reflected the very high level of unemployment on Clydeside, the magnitude of the threat to the 8,500 employees of UCS, and the recognition by the stewards that the political situation was distinctly unpromising.

An Advisory Group of four businessmen appointed by the Government reported on 29 July, recommending that shipbuilding be continued at Govan and that the other two yards be disposed of. On 30 July the Shop Stewards' Co-ordinating Committee announced that the workers had taken over the yards and that a work-in had begun which would only be ended when the future of all four yards was assured. In October 1972 this future had been assured, with Scotstoun, Linthouse and Govan united as Govan Shipbuilders Ltd., a government-

owned company estimated to require at least £25m. to achieve
long-run viability,[4] and Clydebank owned and operated by the
Marathon Manufacturing Company which claims a total investment of £38m. in the new venture, with some £12m. in
Government aid. The success of the work-in was universally
recognised. Indeed it is reasonable to attribute to it an impact
much wider than that felt on Clydeside. The reversal of the
Government's policies for financial support for industry, and in
particular the changes in regional policy, both embodied in
the industry Act of 1972, were certainly influenced by the Clydeside campaign, and it is at least plausible to argue that the
work-in of 8,500 in Clydeside was the prime mover for this
change in industrial policy for the whole nation.

Given what has already been said regarding the lack of support amongst shipyard workers for any formal movement for
industrial democracy or workers' control, how is this campaign
and its remarkable success to be explained? And what significance does it have for the future of shipbuilding trade unionism and industrial policy?

The idea of a work-in was put to the stewards at a meeting
on 13 June. In what comes nearest to an official history[5] there
is reference to the idea emerging previously 'in discussions between some of the stewards', and as it was a Communist steward
who proposed the work-in at the formal meeting of all stewards
it seems reasonable to credit the Communist Party and its shop
stewards with the initiative. The clearest rationale of the work-
in tactic has been given not in statements by the shop stewards
but in a pamphlet by an official of the Communist Party

> The problem facing the leaders of the UCS workers was to
> devise a new technique of struggle which would achieve
> their objective, to prevent redundancies and closures, in
> what was bound to be a tough struggle. A strike could
> play into the hands of the employers when they were set on
> closure anyway. A sit-in would have been difficult to main-
> tain for long enough. It would have also given the employers
> a good excuse to attack the workers by arguing that the sit-
> in made it impossible to fulfil any contract and aggravated
> the bankrupt situation. This could have helped the Tories

to alienate public opinion from support of the UCS workers.[6]

It is tempting to seek explanations of this development on Clydeside in the distinctive history of the labour movement there. Clydebank and Scotstoun had been locations there for the early growth of syndicalism in Scotland, with 4,000 of the 10,000 employees at the Singers plant at Clydebank organised in the Industrial Workers of Great Britain, which was also strong at the Albion Motor Works, Scotstoun.[7] Some of the leaders of this movement were active in the Clyde Workers' Committee in the First World War, and the influence from that time, particularly through Communist leaders such as Willie Gallacher and J. R. Campbell, on Clydeside Communism of the 1970s is undoubtedly strong. But the circumstances of the '70s are so different, and revolutionary expectations so dimmed, that it is not surprising that analysis and policy have also changed. In 1919 Gallacher and Campbell wrote, 'Every struggle has revolutionary potentialities which should be exploited to the utmost'; in this context 'revolutionary' meant 'the struggle to oust the capitalists from the control of industry, the struggle for power'.[8] In 1972 the demand for the right to work was seen as 'revolutionary'. As the official history puts it, 'The work-in is not an attempt to establish "workers' control" on a permanent basis. Such a conception would lack all credibility'.[9]

The essence of the work-in was the demand for security of employment, expressed in the very practical form of a contribution of 'wages', from the shop-stewards funds administered by the shop stewards for workers declared redundant but who continued to report at their respective yards. Initially the intention was that these workers would be allocated work by the shop-stewards, in consultation with the relevant managers. This was only observed for a short time, however, although some staff who joined the work-in on this basis continued to work for several months.[10] The main reason for abandoning the original intention was that once declared redundant a man, not in insured employment, was not insured against injury, and this in an industry with a high accident rate. There was also concern that an issue could develop between management and the shop stewards if a redundant worker 'working-in' were to make a major and costly error. Near to the end of the work-in the Chairman of the Co-ordinating Committee explained

> They reported for work normally but did not work normally. We were not going to subsidise the liquidator by providing labour free of charge. From time to time they filled in for their mates. The work-in ensured that these men were not rotting in the Labour exchange.[11]

These redundant men might have been charged with trespass in yards from which they had been dismissed, but it was one of many illustrations of the desire of the Liquidator to avoid any confrontation that no such charge was ever made. By no means all of the workers declared redundant joined—or remained with—the work-in. In October 1971 of 812 made redundant, 390 (48%) were working-in. By June 1972 1,264 had been made redundant but only 177 (14%) were working-in.[12]

The effectiveness of this 'essential element' of the work-in depended in part on the extent to which the existence of this fall-back wage helped—by giving workers a sense of security—to develop and strengthen their unity of purpose. It also depended very much upon its propaganda effect with the labour movement and the public at large and upon the extent to which it was believed that larger numbers of workers were on the payroll of the shop stewards committee, and involved in a process of workers' control of the yards.

The work-in began with the symbolic control of the yard gates at Clydebank, a statement from Jimmy Reid, 'It is now Upper Clyde Shipyard Workers Unlimited', and an official announcement from the stewards, 'We have totally rejected the statements of the Liquidator. We have told him we will try to keep him out. There will be no violence, but he can do his business from his own office in Glasgow, not here. Nothing has been accepted. Nothing has been discussed. This is a new era in British history. The stewards will decide what will be built, whether it will go out of the Clyde, when it is built—everything.'[13] Much, but not all, of this was rhetoric. The leading stewards recognised that unless they reached a working agreement with the Liquidator supplies would soon run out and the work-in would either collapse or have to be transferred into the less attractive tactic of a sit-in type occupation. But much of the media was less discriminating and realistic, with frequent references to 'workers' control', 'revolution', 'chaos', contributing to the 'visible psychological impact' which Mr. Reid recognised must be made on the public.

Apart from its powerful propaganda appeal (which owed much to personalities) the power of the work-in tactic depended upon the characteristics of the industry, and in particular on the fact that when UCS went into liquidation there were fourteen ships being built in the yards, and that it was in the interests of both the creditors and the prospective owners of these ships (who had made instalment payments) that these ships be completed. This was the main basis of the long period of co-existence between the Liquidator and the shop stewards which began very shortly after the initial propaganda shots had been fired. In addition it would appear that if the demands for which the work-in was initiated were won and the yards continued in shipbuilding use, the price which the new users would pay for them would probably exceed that paid by any purchasers with an alternative use in mind, so that the Liquidator might, in proper pursuit of his duties to the creditors, show some sympathetic understanding to the aims of those leading the work-in.

The point at which the workers were able to influence the Liquidator by action or threat of action came when a ship was due for launching or delivery and instalment payments or final payments could be denied to the Liquidator if the workers chose not to co-operate. Quite early in the period of work-in, workers outside the yard whose efforts were required if ships were to leave the Clyde announced that no such movement would take place unless authorised by the UCS leaders. A similar threat of real (as distinct from propaganda) power was made at the outset, when there was talk of moving a keel destined for Govan across the river to Scotstoun to maintain work in that yard and prevent its closure at the end of 1971. In practice such demonstrations of workers' power were never made, because they were never required.

Although there was a basis for coexistence between the Liquidator and the leaders of the work-in, there was also strain. The success of the propaganda element in the campaign led suppliers to question the viability of the enterprise, frightened away some shipowners who might have placed orders with the new Company Govan Shipbuilders, and gave the world at large, and businessmen in particular, a false picture of workers' control on the Clyde. Towards the end of the work-in the Liquidator made public statements to counteract these impressions.

> There has been a widespread misconception of the nature and extent of the work-in, often misquoted as a precedent for quite different industrial action of a totally obstructive or sit-in nature. A number of creditors and others appear to have been given the impression that the whole operation of the shipyards has depended on the work-in, and that the complex legal, financial and technical and practical problems of building ships, the employment and payment of a large number of employees, and the provision and payment for goods and services, has in some way been organised by committee.[14]

> The work-in was a myth, but a very powerful myth. Without it Govan and Linthouse would certainly have continued, and may have been better placed with orders, but Scotstoun would probably have been closed. Marathon, or someone like them, would have almost certainly come in, but the work-in had secured a significant number of jobs.[15]

The characterisation of the work-in as a 'powerful myth' is correct as far as the narrow aspect of the extent to which workers made redundant actually *worked* under the managerial discipline of the shop stewards, but falls short of a proper assessment. The outstanding contribution of the work-in was to focus public attention on the problem of mass redundancy in an already severely underemployed community, and to mobilise a range of pressures which induced the Government to reverse its policies. Internal to ship-building it established a working unity between the different trades and grades of a deeper and more long-lasting kind than had been achieved—or thought possible —before. Towards the end of the period of work-in this unity survived two tests. One was a division of interest which emerged between the workers at Govan, Linthouse and Scotstoun, whose jobs were by then secured and the workers at Clydebank when finality with Marathon had not then been reached. The difference of interest was accentuated by the fact that, because the Clydebank yard was changing its product from ships to oil rigs, workers there were to receive substantial redundancy payments, even although little or no discontinuity of employment was involved, whereas workers in the other yards were not to benefit in this way. Although initially the narrow self-interest of the men at Govan prevailed, a strong reaction and a threat to resign from the leaders re-established the unity of the entire work-force. The second example is the agreement of September 1972 on wages at Govan which eliminated the traditional boilermakers differential. This not only reflects the extent to which the work-in had overcome schismatic tendencies, but also reflects a determination to maintain and cement the new-found unity of purpose. The charge that the work-in was more an exercise in public relations than in industrial relations fails to give credit for this achievement internal to the labour force and the trade unions. It is apt, however, as far as management-labour relations are concerned. With many of the managers coming to regard the Stewards' campaign as the main hope for their own future security, and with the swell of public opinion behind the work-in, it would have been possible to experiment with new forms of management-worker relations. But the leaders of the work-in did not press for experiments in industrial democracy, although a structural framework for a limited degree of participation was established, as illustrated in Figure 1. The leading stewards saw the issue in straightforward 'black and white' terms, with no gradations of industrial democracy between the traditional managerial system and 'workers' control', and no possibility of 'workers' control' short of a socialist order of society.[17] In this context the work-in was neither reformist nor revolutionary, although the official history claims that it has demonstrated 'the latent managerial capacity of the working class' and that the workers 'most certainly can run industry'.[18] It is part of its success that this is widely believed, although the basis for such belief in the

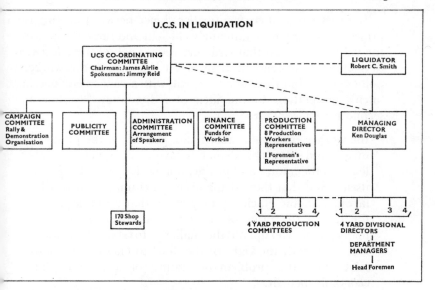

events at Upper Clyde is almost totally absent. The revolutionary context of the work-in lay in the demand for the right to work. In the 1970s even the most Keynsian of economists and politicians may see in this demand a challenge which would have been deflected with greater confidence in the 'fifties and early sixties. R. H. Fryer has posed the issue clearly and in a wider context—

> Where redundancy is in question the choice is between continuing with a blunt variant of the present managerial definition of the problem (as, for example, in making Redundancy payments Act more selective), and giving more serious consideration to the workers' redundancy problem. If the former path is chosen, then the precise form of legislation will depend upon a combination of expedience and appropriateness. If the latter course of action is preferred, then the way forward can be either toward more restrictions upon managerial discretion in the internal labour market and a much more comprehensive strategy in the external labour market[19] or towards a situation in which priority is given to the workers' redundancy problem and the conflict of interests is settled in the workers' favour.

In other words there is a further choice between making the present structure more humanely efficient and radically altering the values on which that structure rests. Each of these last two developments depends initially upon listening to the workers' point of view, but the first does so still with a view to engineering 'men's adaptation to the idea of redundancy',[20] while the second recognises that

> the worker exists in a subordinate position to the property owner and (that) his demands involve an inherent challenge to the nature and justification of the property system itself[21] and that those demands are legitimate. Those who favour humane efficiency may claim that redundancy for the few is the price of economic improvement for the many; those who uphold the challenge to established values recognise both the right of the 'few' to choose for themselves and the problematic nature of any 'universal' economic benefit.

The leaders of the work-in rejected established values and the property system itself but their immediate public aims and ultimate victory were encompassed within the concept of 'humane efficiency'. They faced the recurrent dilemma of revolutionaries in non-revolutionary societies, of how to campaign for reforms not only without damaging the prospects for more radical change, but in ways which would bring the radical change they wished for closer. It is too early to judge whether they have had success in the wider political context which was the main arena of the work-in campaign. It is clear, however, that in the narrow context of industrial democracy little or nothing was achieved. Nothing being thought either possible or important, nothing was attempted.

INDUSTRIAL DEMOCRACY IN A YUGOSLAV SHIPYARD

One observer of the UCS work-in, some four months after it began, claimed that it 'had more in common with the way factories and shipyards operate in Yugoslavia than anything ever experienced in Britain before. It is emphatically not "workers' control" in the sense that the workers actually run

the yards. Working alongside the liquidator is a shadow management, exercised through the yard production committee.'[22] The
committee of eight production workers' representatives and one
foreman representative operated under the jurisdiction of the
Shop Stewards' Co-ordinating Committee, and had access and
links to the top management. But its management function was
more shadow than substance. How much is this also true of self-
management in Yugoslavia?

The extensive literature on workers' self-management in
Yugoslavia does not deal specifically with shipyard production.
Fortunately there is an unpublished study by Frank H. Stephen
of the shipyard at Split which provides some useful insights.[23]
The yard employs 4,500, has four berths and an annual capacity
of 200,000 dead weight tons. There are three levels of Council
through which self-management functions. A 'Collective' or
meeting of all employees can dissolve the top level of Central
Workers' Council by majority vote. The management structure
is circular in form: the workers elect the Central Workers'
Council which in turn elects the Managing Board and together
these instruct the General Director who in turn instructs the
Departmental Heads until the workers receive instructions in
the sheds and on the berths. The relationships between the
management bodies and the executive bodies are illustrated in
Figure II.

The Central Workers' Council has 76 members, each elected
for two years with 38 retiring each year. The number of workers
elected from each department relates to the size of the department. The President of the Council is elected from the Council
for a period of one year. The Council has six Standing Committees:
(1) Socio-economic
(2) Personnel
(3) Protection and Safety
(4) Social Standards
(5) Applications and Complaints
(6) Division of Income.

Membership of Committees is not restricted to the members
of the Workers' Council, but the Chairman is usually a council
member. These chairmen were: (1) an economist; (2) a naval

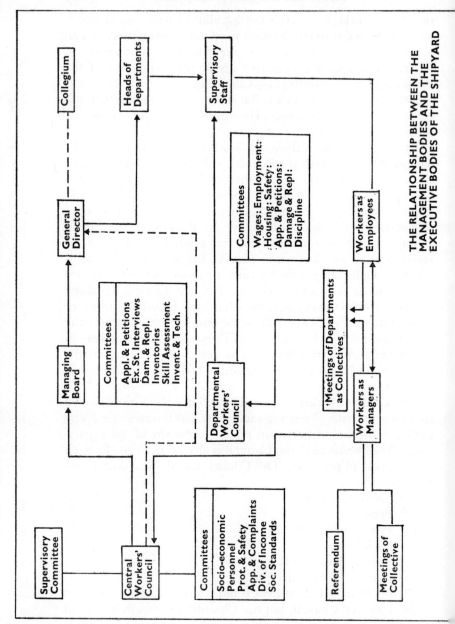

THE RELATIONSHIP BETWEEN THE MANAGEMENT BODIES AND THE EXECUTIVE BODIES OF THE SHIPYARD

architect; (3) a highly skilled worker; (4) a skilled worker; (5) a clerical worker; and (6) an economist. Candidates for the Council are nominated at a meeting of the Collective by a proposer and two seconders, the proposers then speaking to their nominations. The election is conducted by an electoral committee and is by secret ballot at various points throughout the shipyard. In 1967 seventy-six candidates stood for 35 places with every position contested. The occupational composition of the council confirmed Kolaja's[24] hypothesis that the more qualified are selected, but there was an indication of a levelling out as the numbers of unskilled workers and those with only elementary education (working as white collar workers) increased. In discussion with people at the shipyard this imbalance was usually dismissed as the result of the immaturity of the system, the low educational standard of most workers, and the desire to get the best people for the job.

The Managing Board consists of ten elected members and the General Director *ex officio*. The Board is elected for a period of one year by the Central Workers' Council, although not necessarily from it. Workers may be elected for two successive years and are then ineligible for a further two years. This regulation together with a similar one for the Workers' Council, is designed to avoid 'bureaucratisation'. The elected members for 1967 were: two engineers (one of whom is the Chairman), two technicians, one economist, a female lawyer, three highly skilled workers, and one skilled worker. The age range is from 25 to 60 years.

The Supervisory Committee has the job of ensuring that the management of the yard is carried out in accordance with the 'Statut' and the law of the land. There are seven members, each of whom must have been in the industry for at least three years and have a clean disciplinary record. The General Director and members of the Managing Board and the Workers' Council are ineligible for membership of this committee.

Each of the 15 departments in the yard has a Departmental Workers' Council, which is elected at the same time as the Central Council, and on the same basis. The size of these Councils relates to the numbers employed in each department. Workers elected to the Central Council are automatically members of the Departmental Council, and one of their number must be elected its Chairman.

The Central Workers' Council is the supreme organ of management in the shipyard. It is responsible for approving the long- and short-term production plans, the financial plan and the investment plan. Approval of these plans requires a two-thirds majority. However, the procedure by which the plan is drawn up transfers the initiative from the Central Council to the Managing Board and the shipyard planning services. The plan is drawn up by the planning service, then discussed by the Collegium of directors and then passed with their comments to the Managing Board, which makes any alteration it feels necessary. The plan is then posted on notice boards for a period of fifteen days to receive suggestions or criticisms from the Collective. These go before the Managing Board which revises the plan and sends it together with any criticisms or suggestions of the Collective not accepted to the Central Workers' Council. The Workers' Council can then make its own alterations; the final plan passed by it can only be changed at a later date by a decision of the Central Working Council.

The remuneration of labour was based on a rigid set of rules for the calculation of the points allocated to each worker. Each section and skill level was allocated points. Extra points could be given to trades of which there was a shortage to attract workers from other enterprises and also to encourage apprentices to this trade.

Once costs of production and taxes on the enterprise have been met the Council allocates the surplus between various funds (investment, welfare, reserves) and to additional personal income. The councils' role in income determination thus comes at two stages, first when it fixes the value of the point used in the rating system, and second when it decides the share of surplus to go as an additional personal income.

Management discusses almost the same items as the Council but its decisions in the more important spheres can only take the form of recommendations to the Council and all but the most routine decisions are subject to the Council's approval. The General Director and the directors of the different sectors of the yard play a major part in the Board's discussions. The sectoral directors almost always introduce the business affecting their sectors and their recommendations are accepted, after discussion, in the great majority of cases.

Each of the fifteen Departmental Workers' Councils in the shipyard is responsible for the detailed organisation of work within the department and their autonomy is only infringed if the Central Workers' Council considers the interests of the yard as a whole require it. The Departmental Workers' Councils draw up the plan for the department within the framework of the shipyard plan. It plans the phasing of annual holidays, unpaid leave (for short periods), the manpower needs of the department, retirement, educational leave and 'time off', personal applications of members of the department collective, costing of various jobs, termination of employment (with or without notice) and discipline. A committee of each Departmental Workers' Council is responsible for the election of department heads and foremen for a four-year period. The meetings of the Departmental Workers' Council are open to all members of the department. When important items such as the production plan are to be discussed the Council is convened as a meeting of the departmental collective. If the production norms are achieved within the set time the Departmental Council may authorise the division of the financial saving amongst the members of the department. It can also allocate income for replacing machinery within the department. The departments' representatives on the Central Workers' Council can be mandated by the Departmental Council.

The General Director of the shipyard and the directors of the fifteen departments are appointed for a period of four years by a process of open competition. The qualifications for the post of General Director are laid down in the Constitution of the shipyard and must be adhered to when an appointment is made. His salary is tied by the Constitution to five times that of the lowest paid worker in the yard. His functions are laid down in the Constitution of the yard as: running the shipyard in accordance with the decisions of the Workers' Council—controlling the men and machinery—improving production—the working methods—the safety of the workers—reducing costs. He is open to question by all members of the Management

Councils during meetings.

When any decisions of the Management Councils are illegal or contrary to the constitutions of the yard he must inform the Workers' Council and if the warning is not heeded he must report the situation to the commune (local authority). In these circumstances he can refuse to implement decisions of the Council.

There is provision in the rules that prevents the Council altering a decision of the General Director in the day-to-day running of the yard. Once the broad policy has been decided it is up to the General Director to carry it out in the manner which he thinks most appropriate. The Council's only recourse when it disagrees with his methods is to ask the commune for a decision, or to dismiss the General Director and make a new appointment.

The Collegium, comprising the General Director, the Assistant General Director and the Directors of Finance, Production, Design, Purchasing and Sales, is laid down in the *Statut* not as a management organ, but as a forum of experts. It discusses the problems of the shipyard in an informal manner and seeks the advice of the heads of various departments. The main function of the Collegium is to advise the General Director on technical problems, but he is not bound by the conclusions of the Collegium. Any of the members can ask to have his opinion minuted, so that if the General Director ignores advice the position of his expert advisers is safeguarded.

It is clear that machinery exists for the effective control of the shipyard by the workers employed in it. The Yugoslavs include in the term 'worker' everyone employed in the enterprise. 'Workers' control' therefore includes control by elected executive staff. The management of this yard remained with 'managers' and white-collar specialists elected by the workers. The role of the specialist has not been eroded by the system— evidence of this is seen in the fact that the production-technical committees of the Council are chaired by the appropriate specialists.

The Organs of Management in the shipyard provide an excellent structure for democratic management—a large 'general' council and smaller 'executive' committee. There is complete freedom for the workers to choose the representatives they wish and the voting figures show that this freedom is utilised. The trend towards a greater degree of decentralisation of management may give the workers a greater degree of self-confidence, since it will bring a greater number into contact with the decision-making machinery. What would seem to be lacking in the confrontations between the executive management and the ordinary members is alternative policies and counter proposals. The executive makes proposals and these are either approved or referred back for reconsideration. The initiative is never taken away from the executive.

The existence of external constraints on decisions taken within the shipyard should also be noted, some diminishing and some enhancing its degree of industrial democracy. On the restraining side are the financial checks operated by the district Bank, which controls the credit facilities so necessary to ship-building, and which also rations foreign currency. The Bank also credits the shipyard's books and deducts the appropriate taxes and social insurance payments. Encouraging the development of workplace democracy are the shipyard branch of the trade union and the local Commune. The trade union is concerned to ensure that the Workers' Council functions effectively and that the decisions of the several management committees are fully carried out. The Commune has some statutory responsibilities for overseeing the operation of the shipyard, encourages reluctant or apathetic workers to involve themselves in the processes of management, and keeps a watchful eye on the dangers of bureaucratisation developing within the shipyard.

Returning to the question posed at the beginning of this illustration from Yugoslavia, the degree of participative democracy in the running of this shipyard is substantial. Both the framework and the desire for the further development of industrial democracy exists. With all the weaknesses and limitations which have been noted in the general literature on workers' self-management in Yugoslavia, the situation which Stephen describes in the shipyard at Split confirms in that particular example the generalisation made of the system by Riddell:

a large-scale attempt to decentralize the control of social institutions is an unique response on a national basis to problems of organization and development in a modern industrial society, although one based on conceptions that have interested the labour movements of many other countries.[25]

The impact of self-management on even the most important policy decisions affecting the shipyard was illustrated when in 1967 the Federal Government recommended that the yard at Split be merged with three other independent yards to form a consortium. At Split 18.3% of the workforce (21.3% of those voting), voted against the merger. At one of the other yards— the Uljanik yard at Pula—the majority were against. After six months in which there was further explanation of the benefits expected from the merger the merger was endorsed by majority vote in each yard. This experience contrasts sharply with that on Clydeside before the formation of UCS, when it is doubtful whether the workers at Fairfields, at least, would have voted to enter the new consortium had their opinion been canvassed.

POSSIBLE FUTURES

Against the background of the general revival of interest in industrial democracy and the particular experiences of ship-building workers on Clydeside since 1966 it is interesting to speculate on the possible developments in the industry. This background has led to possible suggestions from the top-level trade union organisations within the industry which, though certainly modest, reflect a new interest in 'participation'.

If trade union leaders and workpeople are expected to accept the heavy responsibilities for the industry which management, and the Government, continually press on them they should be given the right to participate effectively on behalf of their members at every level in the industry, from individual yards to the proposed National Shipbuild-ing Council. Such a body should regard as one of its main functions the creation of an effective system of yard and district planning and production committee designed to win the interest and co-operation of both local management and workpeople and draw on their practical experience. In this way energy and enthusiasm which are now dissipated or untapped could be applied to the task of reforming the industry.[26]

The extent to which industrial democracy is developed in shipbuilding will depend on whether the industry remains in the private sector, or if it is taken into the public sector, the organisational structure which is adopted. Within a private sector, the Fairfields experiment demonstrated that some extension of industrial democracy was possible, partly by structural changes which created a new representative role within management for the stewards, and partly through enhanced trade union influence within productivity bargaining. A conscious attempt to develop single-channel bargaining at shipyard level, combining the consultative and negotiating procedures, coupled with some form of representation within management of the Fairfields type is probably as far as any development of industrial democracy could go within a shipyard in private (or quasi-private) ownership. Such a development would mark a very considerable extension in participation on the present situation in British shipyards. Within such a framework it is to be expected that the informal controls over work processes which individuals and groups of workers have established in shipbuilding would be brought out into the open, formalised and encouraged to operate constructively rather than restrictively.

If shipbuilding were taken into public ownership, industrial democracy would have greater opportunity to develop if the structure were a decentralised one, with a substantial range of powers and decisions lying with the managements of individual shipyards. The production characteristics of shipbuilding would also favour such a decentralised structure.

Before Government provided financial support for the new Companies which took over the old UCS yards there had been suggestions of municipal and other forms of ownership, including the (financially impractical) proposal that the trade unions buy and run the yards. J. E. Meade has observed that 'easy company promotion—by unemployed or ill-paid workers will certainly be impossible without appropriate governmental interventions of a most extensive character—leading perhaps inevitably to a socialist ownership of the main capital resources of the community as in Yugoslavia.'[27] It is interesting to speculate on whether the creation of such a 'labour partnership in one or more of the Clydeside yards would have transformed the attitudes of the workers to the linked questions of redundancy and manning. Several economists have demonstrated that such a labour partnership would seek to maximise income per head of numbers employed, and that with decreasing returns to additional labour input there would be a tendency to favour a cutback in employment to increase income per head.[28] Whether such a policy would be adopted and implemented would depend upon a number of factors, of which the influence of the mass vote or collective is one, the proportion of workers whose jobs could be put at risk is another, and the extent to which each worker would know whether his own job security would be at risk or not is a third. With overmanning as a major problem in British shipbuilding, and given the strength of worker resistance to redundancies, it would have been interesting had this theory been put to the test. In the Yugoslav context Vanek has argued that this Ward/Domar model is inappropriate, and that the Yugoslav enterprise has a number of goals.[29] It seems likely that a 'labour partnership' on Clydeside would also have been a 'satisficer' rather than a 'maximiser' aiming only at optimising income per head of those employed.

Whatever the form of public ownership it is unlikely, in the early stages at least, that a structure of democratic committees with powers to select managers such as exists in Yugoslavia would be adopted in Britain. There are two reasons for this. The first is that with a stronger, independent trade union movement in Britain there would be a preference for achieving a higher proportion of any shift to industrial democracy by extending bargaining rights over a wider range of issues at the expense of more direct democratic forms. The second is that such a major shift to a system in which executive management functions within a committee structure based on elective procedures would only be practical (because acceptable) if such a system were universal or at least widely adopted. It could only come, that is, in the wake of a major political change rather than encompassed within a change of the ownership of a particular industry.

Writing at the beginning of 1973 little change over the rest of this decade seems more likely than much. Remembering, however, the experimental temper and dramatic events on Clydeside which would not have been expected or predicted seven years ago it is probably wiser to forgo any prediction of which if any of the possible patterns for industrial democracy suggested above will be adopted in British shipbuilding.

REFERENCES

1. G. D. H. Cole, 'Engineering & Shipbuilding', *Chaos & Order in Industry*, Methuen, London, 1920.
2. A full description and analysis of this experiment will be found in K. J. W. Alexander and C. L. Jenkin, *Fairfields: A Study of Industrial Change*, Allen Lane, London, 1970.
3. House of Commons, *Debates*, Vol. 722, Col. 2103, 22 December 1965.
4. *Shipbuilding on the Upper Clyde, Report of Hill Samuel & Co. Ltd.* Cmnd. 4918, HMSO, London 1972. See especially Notes to Appendix 11, p. 20.
5. Willie Thompson and Finlay Hart, *The UCS Work-in*, Lawrence & Wishart, London, 1972. In his foreword Jimmy Reid, the official spokesman for the Co-ordinating Committee, writes, 'My belief is that their analysis is by and large a correct one. The future may require that judgements made be modified, but I am certain that these will be modifications, not re-evaluations.'
6. A. Murray, *UCS—The Fight for the Right to Work*, Communist Party, London, 1971.
7. Thomas Bell, *Pioneering Days*, Lawrence & Wishart, London, 1941.
8. Wm. Gallacher and J. R. Campbell, *Direct Action: An outline of workshop & social organisation*, 1919. Republished as Reprints in Labour History, No. 3, Pluto Press, London, 1972.

9. Thompson and Hart, op. cit., p. 94.

10. *Report by the Official Liquidator to the Creditors of the Company for the year ended 14 June, 1972*, Glasgow, 28 September 1972.

11. Iam Imrie, 'How men and management view work-in', *Glasgow Herald*, 28 September 1972.

12. *Report by the Official Liquidator*, op. cit., Appendix IV.

13. *Glasgow Herald*, 31 July 1972.

14. *Report by the Official Liquidator*, op. cit., p. 11.

15. Mr. Robert C. Smith, the Liquidator, in a speech to the Publicity Club of Glasgow, reported in *The Scotsman*, 18 August 1972.

16. A. Hargrave, 'How they're managing at UCS', *Financial Times*, 27 October 1971.

17. This judgement is based on a number of sources and in particular on statements made by Mr. James Reid at a Conference of the Society of Industrial Tutors held at the University of Strathclyde on 14 to 16 April 1972.

18. Thompson and Hart, op. cit., pp. 91–2.

19. Robert H. Fryer, *Redundancy, Values & Public Policy*, Discussion Paper, Industrial Relations Research Unit, University of Warwick, 1972.

20. W. W. Daniel, *Strategies for the Displaced Worker*, London, P.E.P., 1970, p. 10.

21. Daniel Bell, 'Industrial Conflict and Public Opinion', in A. Kornhauser *et al.* (eds.), *Industrial Conflict*, McGraw-Hill, New York, 1954, p. 242.

22. Hargrave, op. cit.

23. Frank H. Stephen, *Management Structure & Industrial Relations in a Yugoslav Shipyard*, mimeo. Glasgow 1967. Mr. Stephen has kindly permitted the author to draw extensively on this study.

24. J. Kolaja, 'A Yugoslav Workers' Council', *Human Organisation*, Vol. 20, pp. 27–31.

25. David Riddell, 'Social Self-government: theory & practice in Yugoslavia', *British Journal of Sociology*, Vol. 19, 1968.

26. *The Shipbuilding & Ship repairing Industry*, a mimeographed policy discussion document prepared by the Confederation of Shipbuilding & Engineering Unions, 1972.

27. J. E. Meade, 'The Theory of Labour-Managed Firms and

of Profit-Sharing', *Economic Journal*, Vol. 82, no. 325s, 1972.

28. B. Ward, 'The Firm in Illyria: Market Syndicalism', *American Economic Review*, Vol. 48, 1958, and *The Socialist Economy*, Random House, New York, 1967; E. Domar, 'The Soviet Collective Farm as a Producer Co-operative', *American Economic Review*, Vol. 56, 1966; J. Vanek, *The General Theory of Labour-Managed Economics*, Cornell University Press, Ithaca, 1970; Meade, op. cit.

29. J. Vanek in 'Discussion' in M. J. Brockmeyer, *Yugoslav Workers' Self-management*, Reidel, 1970, p. 167.

Workers' Participation in Private Enterprise Organisations

GEORGE F. THOMASON

AUTHORITY AND PARTICIPATION

There are many different forms of private enterprise organisation, from the classical entrepreneurial concern through the partnership and the private company to the public joint stock company—many of which now have international associations. All of them have in common that they distribute authority to decide on the principle of the primacy of ownership. In spite of the variety of form, and although, even in the archetypal joint-stock company, there are now necessary qualifications to be made to the applicability of the ownership principle, the distinguishing characteristic of private enterprise is the use of ownership to legitimate the exercise of authority.

In the simple entrepreneurial concern, the exercise of authority by 'the owner' is clear to see: the entrepreneur can be seen to decide all matters of moment. In the greater complexity of the modern joint stock company the discovery of just who takes the decisions about what is a more difficult task. Similar difficulties are encountered in the nationalised industry or the large co-operative society. But if the problem is associated with size and complexity, the solution to it may well lie with certain categorisations of decision which have now become fairly well accepted.

In a quite early work on this general subject, Chamberlain discussed the challenge of the trade unions to management control in terms of three broad *levels* of management decision. These are capable of providing us with all the schema that we need to organise thinking about 'participation' (Chamberlain, 1948).

The first category may be referred to as *directive* decisions, entailing the use of final authority—i.e. authority which cannot be overturned except by another decision by the same person or body—to determine the corporate purposes and objectives and the overall policies. In a private company this authority is associated with the Board of Directors.

The second category centres on *administrative* decisions, concerned with *means* of implementing these overall policies in the pursuit of the given objectives; the departmental and higher middle management is primarily concerned with such issues and they are vested with delegated authority to enable them to take these decisions. Lower middle management and supervision carry out the function of 'implementing' these decisions, this process requiring *executive* decisions to secure *fit* to the circumstances. Such hierarchy of directive, administrative and executive decision levels is to be found in organisations other than those to be categorised as private enterprise, but the fact of the existence of such hierarchy in the latter type of organisation enables us to consider the various manifestations of participation to be found there.

The term 'participation' is also susceptible to a variety of meanings. French's definition of this term is probably the most useful and sensible for general purposes. 'Participation refers to a process in which two or more parties influence each other in making certain plans, policies, and decisions. It is restricted to decisions that have further effects on all those making the decision and on those represented by them' (French, Israel and Aas, 1960). For the present purpose, it will be employed to indicate a process of sharing, not merely in economic benefits of enterprise, but also in the process of decision-taking about the means of achieving such benefits.

Discussion of participation in private enterprise organisations must require a restriction upon the meaning of 'workers' participation' as employed in certain other contexts. Given the association there of ultimate authority with ownership, workers' participation in private companies cannot mean what the Marxists, the syndicalists and even the guild socialists mean by *workers' control*; these are essentially 'replacement' ideologies, i.e. supporting a replacement of the owners as the power elite by a worker or worker-plus-state elite. For present purposes, therefore, we must define workers' participation as a process in which workers share or take some part in the decision-making function of the enterprise, and do so as of right or at least on a basis other than the accidental or quixotic.

Other chapters have dealt with the question of such worker sharing in decision-taking within enterprises which either were ceasing or had ceased to be 'private' in the usual sense of the term. Here we are concerned rather with those discernible developments in organisations which have not ceased and are not ceasing to be recognisable as private enterprise concerns. Where it is possible to argue that both nationalised (and municipalised) industries and the producer co-operatives entailed a modification of the rights of owners to decide about their own and an alteration of the relative power of the workers in the system, it is also possible to claim that these changes in effect eclipsed private enterprise in its usual sense. Nevertheless, there are private enterprise concerns which have also adopted devices in many or most respects comparable to those of the nationalised industry or the co-operative society, but which are still recognisable as 'private' concerns. These are our present concern.

The advocate of replacement structures will, no doubt, regard this area of 'participation' as irrelevant, representing a palliative or sop, and as a device to delay the real development of workers' (or some other category's) control. Nevertheless, in our mixed economy a large proportion of workers are still employed in private enterprise companies, and there are not very visible proposals or plans to change this, even if there are proposals to alter the structure of decision-making within them. These proposals are to be found in the last of the Labour Government's consultative documents which proposed 'participation' for private companies on the basis of some modified form of either the British Steel Corporation's Employee-Director's scheme or the co-determination arrangements perpetrated or proposed in some of those continental European countries which form partners with us in the enlarged European Economic Community. Whatever the 'ultimate' outcome of these changes, therefore, in the short run, we are more likely to see developments in the direction of shared decision rather than replacement of the decision-takers.

It can indeed be argued that the term participation has been invented to admit of just this kind of change; to consider the other types, the older concepts of workers' control or industrial democracy could well suffice. In Walker's (1970) treatment of the subject moreover, he both refers to this kind of sharing in decision-authority and employs a categorisation which depends upon the kind of distinction as to level which has been presented above. In this statement on the subject he identifies:

(a) participation in ownership;
(b) participation in government;
(c) participation in management; and
(d) participation in setting the terms and conditions of employment.

The last of these might best be described as a participation in a type of decision rather than in a level of decision; the implication is that because we are primarily concerned with 'worker' participation, then the determination of the terms and conditions under which these workers will work provides an area of decision in which they might be expected, quite logically, to have a special interest regardless of the 'level' at which such decisions might be made in the organisation.

The other three of Walker's categories might, however, be associated rather more directly with 'level' of decision. The first of them is, perhaps, least easy to link in this fashion. It is primarily concerned with shareholding (at least within the context of our kind of private capitalism), and therefore one might presume with the sharing of the *benefits* associated with ownership. Whilst these may be thought of as essentially economic benefits (Parkinson, 1951) we should not lose sight of the benefit of 'ultimate authority' in a private enterprise venture, even if the modern corporation is beginning to assume a high degree of independence of its shareholders in annual general meeting assembled (Brown, 1953). Just because ownership allows a benefit of power or authority in our society, in addition to the benefit of a share of the 'goodies', we would be justified in looking at participation in ownership as a means of sharing 'ultimate authority'. Both the syndicalist and the nationaliser would have this interest in ownership.

Since, however, not all decisions which are of important concern to people in organisations are taken by those who possess this ultimate authority, we can also consider the issue of sharing in relation to these. There is, as Hughes (1965) amongst others has suggested, a hierarchy of decision, and even if it is unnecessary for us to categorise this in as fine a detail as either Paterson (1972) or Jaques (1956), it could be useful to look at this 'lower-level' participation in decision taking. To do so, however, we need to make no more detailed a distinction than that used by Chamberlain to separate the *administrative* decisions about means of realising objectives and implementing overall policies and the *executive* decisions concerned with the fitting of pre-determined programmes of action to the situation in which that action occurs. Neither of these is directly associated with the exercise of 'ultimate authority' although both rely upon it; together they identify the category of 'participation in management' as defined by Walker.

	Ideology Commitment to Private Enter- prise	Challenging of Private Enterprise
Decision Level Overall Objectives and policies Exercising Ultimate Authority	Co-determination & Commonwealth experiments	Replacement & with- drawal strategies
Administrative and Executive Decisions Exercising Delegated Authority	Joint consultation & Joint Committees	Bargaining relation- ships

Essentially, therefore, we can distinguish four broad categories of participation in the private enterprise framework which derive from a juxtaposition of varying ideologies relevant to work organisation and the levels of decision within private enterprise organisations. The one entails a distinction between the ideology of private enterprise and the ideology of some other actually or potentially challenging system of determining the allocation of 'ultimate authority'. The other—for present purposes—distinguishes the ultimate decision from the consequential decision, i.e. the decisions associated with ownership and ultimate authority from those often referred to as administrative and executive. Brought together in the form of a matrix, these yield a four-fold categorisation, although for the present purpose, the two ideological varieties of participation in administrative/executive problem-solving will be treated as one.

Although a distinction can be made at the administrative-executive level, as in the table, it seems sensible in the present set of circumstances to examine the varieties of participation at this level under one general heading, if only because the ideological distinctions at this level are difficult to disentangle without specific research activity in each particular case. At the 'higher' level, the ideological distinction is worth maintaining in the discussion because it is clearer and because it is associated more obviously with different strategies and outcomes.

PARTICIPATION THROUGH JOINT PROBLEM SOLVING

The first discernible category of participation is that which focuses upon a sharing of authority to decide administrative (Chamberlain, 1948), conditional (Jaques, 1951) or consequential questions. These may be defined as those questions which remain for determination even after the corporate objectives and policies have been established, and which have been described as the traditional prerogative of 'administrative' management or concerned with decisions as to 'how?' the objectives and the policies are to be implemented. As in the constitution of joint-consultation in the Glacier Metal Company Limited (Jaques, 1951) participation at this level can well leave untouched the more fundamental questions which lie within the prerogatives of the Board.

It may be asserted fairly confidently that most of the experiments with new (joint) decision structures in British Industry have occurred at this level. They have not been concerned to change the 'ultimate' authority structure, but rather to take this as a given in their search for solutions of those consequential day-to-day problems which remain for management and men to resolve. A great deal (although not all) of joint consultation has been focused in this way, and so has a good deal of the more recent productivity bargaining in which managements took the proffered opportunities to move bargaining on to a new plane of joint problem-solving. In fact, much of the reported development of 'joint' approaches to this and that has been concerned with just this kind and level of decision-taking (Thomason, 1971). In Walker's categorisation, this is participation in the *management* decision process (Walker, 1970).

In this area of decision, it is possible to distinguish a 'three-pronged' attack on the problem, in which communication precedes involvement, which it is hoped will in turn lead on to commitment. This is clearly exemplified in the John Lewis Partnership or the Avon Rubber Company approaches for example. But it has fallen traditionally to the lot of an institutionalised joint consultation to solve the first two parts of the problem, and, hopefully, to lead into the third; more recent experiments have followed on this line, but with perhaps more sophisticated approaches and techniques.

(a) *Joint Consultation*

The topic of joint consultation has, over the years, attracted a voluminous literature to itself. It is usually considered to have begun in the First World War period (helped by the growth of the shop stewards' movement and the Whitley Committee recommendations), to have dropped away in the inter-war period in the face of depression, to have revived with Ernest Bevin's joint production committees of the Second World War, and then to have fallen away again only to become caught up in the post-war drives for (firstly) better human relations and improved communications and (later) increased employee motivation in the face of developing alienation or apathy. This chronological background is important, but the subject of joint consultation cannot be considered in isolation from developments in joint regulation (or collective bargaining, to use the more common term). Nevertheless, the object of this section is not to repeat yet again the history of joint consultation, but to link the practice with current conceptions of 'participation'.

From this point of view, the essence of joint consultation lies in the opportunity which this device presents to secure discussion of 'common' problems of life and work in the work situation. In distinction from collective bargaining, joint consultation either assumes or seeks to develop consensus, or at least a sufficient degree of it to permit meaningful or fruitful discussion of some issues which arise in a work organisation. It is now, of course, common enough to hear complete denials of any common interest or consensus in industry—particularly where the subject of participation is under discussion. A more realistic view might be that any system of on-going relationships will contain and even depend upon both conflict and consensus, both common and separate interests; it is therefore as possible to find something which all parties will be willing to discuss as to find issues which will divide them. The important question to be resolved in the particular case is just where the dividing line is to be drawn—and the opposed claims that joint consultation is never concerned with anything more important than canteen tea and that it sustains a whole factory culture seem to imply that it can be drawn at widely separate points on the spectrum. There is little doubt that discussion in this framework, as the radical trade unionist is liable to claim, is intended to develop, and may well succeed in creating, greater consensus within the enterprise, but this could be only slightly less true of discussion within the framework of collective bargaining.

When joint consultation exists in distinction from the collective bargaining machinery—which has been the common position hitherto—it does generally function as a means of joint discussion of problems in which both or all parties have some common interest. In ordinary, day-to-day conditions, these problems may well be no more world-shattering than those associated with amenity, e.g. canteens, car parks or bicycle sheds, or with safety, a subject on which the various parties cannot let it appear that they are opposed to it. In emergency conditions, such as a war or (perhaps) stagnation in the economy as a whole, a presumption is created that all men of goodwill must have an interest in the levels of production attained, so that 'last month's production' becomes an acceptable topic for discussion; in this connection, as with the safety question, joint consultation can be employed as a means of legitimising exhortation to greater effort or contribution, but usually at a verbal, non-sanctioned level. On the other side of this coin, when the condition is one of threat to the basic securities of the employees (as, for example, in the face of a possible plant closure and consequent redundancy) the worker representatives may similarly find the consultative arrangements a useful and usable device for securing more information from the management on 'the position' (Fox, 1965).

Joint consultation thus involves participation in discussion. This discussion may result in decision in some cases, but it has in the past, most often, produced decisions which were acceptable to the management but which were probably not capable of being imposed upon the workforce in a simple autocratic fashion. It is, for example, difficult to impose decisions about safety without prior consideration or 'softening up'. Furthermore, such decisions as do tend to emerge from this arrangement could be described as decisions within the purview or jurisdiction of the administrative management (in the sense in which Chamberlain defines this term 'administration'); they do not, in other words, seriously challenge the final authority of the management, but rather smooth the passage of decisions about *how* certain given objectives might be achieved.

These conclusions about run-of-the-mill consultation are well illustrated in what has probably been the most elaborate exercise of this sort to secure widespread and detailed publicity, that of the Glacier Metal Company, reported extensively by Jaques.

The Glacier Model

In this case, consultation began in this fairly usual fashion, as a consequence of wartime persuasion to establish factory production committees. Under Wilfred Brown's direction, and with the assistance of the Tavistock Institute personnel, this was refined and developed into an elaborate structure of committees which permitted many different groups to engage in consultation at different levels in the hierarchy as well as at the level of the factory as a whole. During the process of 'working through' the problem of consultation, the Company evolved two main documents which sought to provide a constitution for the various committees, the most important of which was the Company Policy Statement.

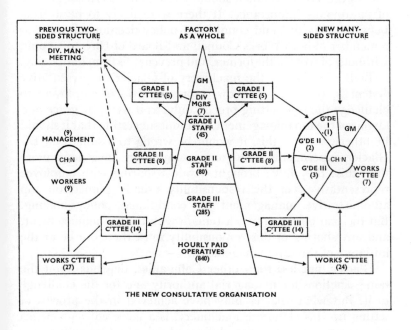

THE NEW CONSULTATIVE ORGANISATION

This lays down clearly what policy subjects can be discussed and decided through the Works Council on which sit representatives of all the different grades, seven from hourly-paid grades, and three, two and one representatives from grades III, II, and I staff, with Brown as Chairman. A distinction is made between that *definitive* policy which must continue to be made by the Board of Directors as having a statutory responsibility in this matter, and that policy, *conditional* upon this, which can be made by a body elected from the employees of the establishment.

The most important feature of the Glacier set-up is the attempt to provide employee participation in rule setting whilst yet recognising that a representative body drawn from employees cannot make *all* the rules, because the law would not allow it. Therefore, the requirement in Glacier that any decision of the Works Council should command a *unanimous* vote is of paramount importance. If there were likely to be a clash between definitive and conditional policy decisions, Brown, as a member of both Works Council and Board of Directors could withhold his vote in the former, and prevent this from happening.

Elaborate though the machinery of Glacier's representative system is, therefore, it allows representatives of employees to challenge the rule setting process only, as it were, by grace and favour of the top management. Without suggesting that it should be otherwise than this, it would be wrong to assume that the Glacier experiments with joint consultation have done more than this: they have brought about a sharing with employee representatives, of the responsibility usually pinned on top administrative management for conditional policy making. But the real power must yet remain with the top management, and any sharing of this responsibility must therefore be at the initiative of the top management.

Glacier, not less than other Companies, depends upon the same sanctions for managerial authority and for the challenge to it. But what Glacier may have achieved in the process of setting up this elaborate machinery, is a more efficient means of communication between different levels of a hierarchy and a greater degree of involvement, without fundamentally altering the basis of the authority in the hierarchical organisation. By providing numerous opportunities for discussion of policy problems, both at the different levels and by Department, and overall in the works council, the possibility of achieving greater consent to the administrative rules, would seem to exist.

This function of improving communications to the end of securing greater consent is a common justification for the formal creations of joint-consultative machinery. Most often this starts from the assumption that the need for communication and consent is greater at the hourly-paid staff level, and top management therefore tends to develop a formal means of communicating directly. As in Glacier, however, this usually leads to a 'squeezing out' of the middle management layer, resulting in some resentment, and some deliberate arrangement has to be made to bring this group back into the structure of top-to-bottom communication. As Brown argues, this ought not to happen if the real nature of the representative system is recognised, but in practice it does (Brown, 1960). For it is one thing for Brown to say that a manager ought not to criticise representatives for representing views which he disagrees with as an executive; it is a much more difficult thing for them to do this in practice. It is one thing to say that the shop steward should leave outside the committee room notions of class warfare as irrelevant to the issues to be discussed in the committee, and it is another to achieve this result in practice. Neither are impossible of achievement, but both are difficult.

This centres on the problem of role-distinction. A man working on a bench, or a manager working as a manager, has one role to perform within what Brown would call the executive system—the job of carrying out whatever instructions he has received. But when either of these assumes the role of a representative in relation to joint consultation, he is not any longer carrying out this role, and the definitions appropriate to this role no longer apply. Each has really a different job specification for that part of the time when he is acting as a representative. But it is difficult for individuals to slip out of one role and into a completely different one when the basic situation and the members of the group do not really change. This same problem also arises for the man who is not only a part of the executive system, but also a representative within a trade union framework, *and* a representative of a constituency in the works council.

(b) *The Plant Bargain Model*

It happens often that this last kind of situation develops out of a joint consultative approach. Both in Glacier and, indeed, in many other experiments with joint consultation initiated by the management, the original position of constituency representatives being non-unionists elected by non-unionists (or partly so) often proves extremely difficult to hold, and union representatives often take over these representative roles in addition to those they occupy within the union framework.

But more than this, it can also happen that the official trade union machinery of representation and bargaining becomes part and parcel of this same functional type of joint consultation. Although founded in conflict of interest, it is possible for the local trade union representatives and the local management to develop a joint problem-solving approach usually within the framework of 'plant- or company-wide bargaining'. Whilst it is true that behind this there is the panoply of trade union sanctions, it frequently proceeds as if such sanctions were not present.

This type of development may be illustrated from the Halesowen Foundry of GKN-Shotton whose Charter for the Workpeople provides for a Factory and Productivity Council and four sub-committees through which the local problem-solving is processed. The Charter was introduced after a three-month strike of predominantly immigrant workers in the Foundry in 1969, when the management dismissed the strikers and recruited a new work-force less predominantly immigrant than before. It provides for a sophisticated joint regulation structure involving the shop stewards and convenor, and has had interesting consequences since it was introduced.

The main Council comprises eight management representatives and eight worker representatives and meets three times a year under a Management Chairman whose place is taken by a worker representative if he is absent. The function of this Council is to review the present and future performance of the company with particular attention to developments in manufacturing capacity, volume of employment, training and redeployment, to consider ways of maintaining employee interest in the company and its methods of working and to prepare and review productivity bargaining within the company. The Council also reviews the work of four consultant sub-committees which are concerned with:

(a) agreements and factory manning;
(b) safety, health and welfare;
(c) employment benefits and leisure activities;
(d) discipline.

The first of these committees meets not less than two-monthly and is made up of three management and three shop representatives. It is charged with the negotiation of any changes in terms and conditions of employment, making procedural agreements on any matters affecting these, planning productivity agreements for recommendation to the Council, and to offer advice on manpower budgeting and training. It is supplied with information on future orders, production and loading, and other marketing and production matters to enable the trade union members to discuss matters intelligently—and also to form a shrewd estimate of profitability even though financial information is not given.

The Safety Health and Welfare Committee is composed of five members, two being shop representatives and two management nominees, together with a representative of staff workers. It is the main function of this Committee to review and advise on the administration of all matters relating to safety, health and welfare in its two-monthly meetings. The employee benefits and leisure activities committee comprises seven members split in a similar way to the safety committee and is charged with reviewing and advising on the administration of such employee services as pensions, sickness benefits, benevolent grants and charitable grants, and with establishing a company suggestion scheme. It also examines existing sporting, social, educational and leisure arts facilities provided for employees and largely administers these.

The Discipline Committee comprises four members drawn from shop representatives and management nominees, and has the responsibility to ensure equity and just treatment of employees in cases where warning, reprimand, suspension or dismissal have been necessary, and for hearing appeals against such decisions. The Committee also considers and agrees the works rules and recommends lines of action for the maintenance and observance of these rules. It meets as necessary particularly in cases involving indiscipline or disciplinary action considered necessary by management. In its review of such cases it has power to suspend or dismiss employees.

It has in fact suspended a few workers for lateness and dealt with two others who fell out with one another, but it has not yet dismissed anyone. Cases dealt with are few, and it is reported that between February and October 1972, no discipline cases were brought before it—regarded as a testimony of self-discipline in the Foundry under the new arrangements.

The advantages claimed for the Charter are of two kinds. Firstly, the production has increased (tonnage output has doubled and scrap rates have been cut from 40 to 14 per cent, whilst the workforce has dropped by fifty to the present three hundred and labour turnover from 40 per cent to 4 per cent) and costs have fallen although earnings have moved from £24 which included £6 tonnage bonus, to a possible £40 including £20 tonnage bonus with less overtime working. Secondly, it is claimed that the industrial relations climate has improved considerably. The evidence produced for this includes better team-working amongst different ethnic groups and a more constructive approach to the solution of 'common problems' surrounding manning, etc. But the works convenor has been reported as saying, 'You never hear the word "strike" in this company. We're given the information to know how the company is doing, and we're pretty certain that our wages will go up with the prosperity of the company.' The Personnel Manager has commented that 'the charter is a lot of words that are meaning-less without good will and good communications' and the Company considers that it now has both. But the implication is that this arrangement has moved the originally conflict-based structure into a framework of co-operation to solve the common day-to-day problems, so that the organisation becomes func-tionally indistinguishable from the better or more sophisticated joint consultative arrangements.

The features exhibited by the GKN-Shotton case are becom-ing increasingly common in major industrial concerns. The criteria of the NBPI in the second half of the 1960s provided an opportunity which a number of Companies took, to develop a 'productivity bargaining approach' to their manning and work allocation problems which emphasised the discussion, involve-ment, commitment sequence. In decision-taking terms, those bargains which owed something to the Oldfield method, and some of the others, sought and served to bring to the bargaining table issues and constraints which do not normally secure much attention there. Amongst these are the influences of national and international market changes upon the local situation. Such factors whilst often introduced by management in bargaining, are often treated by the other side as 'management's problems' to resolve.

Companies like BP Plastics Department and Associated British Tissues do however claim modest success in getting people on the shop floor to think about earnings rates in relation to the Company's competitive position, and to think about flexibility less in terms of the shopping list and more in terms of overall contribution to efficiency. To do this, it has been necessary to invest a good deal of time in cascading discussions of productivity and efficiency, and although this does not by any means work out 'perfectly', they have found that both the resultant suggestions and the consequent action reflects this growing knowledge and awareness of the constraint pattern.

These experiments also seem to produce an acceptable pay-off. In the BP Plastics' case, for example, average hours of work dropped from over 48 to $42\frac{1}{2}$ and economies were secured by a greater flexibility of working as between craft and process. This enabled the employees to secure a number of benefits associated with staff status, but it also indicated to management that this more open participative style was worth the extra time and effort involved in its development.

There seems, however, to be little point in going into this exercise unless the management is willing to do two things: firstly, to listen to suggestions, no matter what their content (e.g. whether labour-saving or capital-requiring), and to spend the necessary time in discussing them *realistically*; secondly, to provide training or opportunity for training *to the level of skill* which is required for participation in the decisions which management is thus putting on the bargaining table.

It is easy to turn down suggestions which require *capital investment*, but to do so without discussion is to miss the opportunity to discuss the constraints upon management decisions. It is also easy to limit training to a few simple 'talks on' something or other which is considered relevant, but this misses the opportunity to improve the level of *general* competence amongst the labour force. As Richbell found during an evaluation of a detailed productivity bargain, the amount of flexibility achieved may well prove to be a function of the amount of training given.

The solution of the problem of 'indiscipline' in industry

might well lie in a similar sharing of management's power to determine whether the rules have been infringed. Not only does this happen in the GKN-Shotton case. It has also been attempted successfully in the Firestone Company, where a Misconduct Committee has been in existence for twenty years to arbitrate upon any breaches of discipline. Consisting of two stewards, one other worker elected from the Works Council, the Departmental Manager and Foreman and a Chairman from the Industrial Relations Department, it adjudicates upon complaints of bad workmanship, insubordination, fighting, time-keeping, etc., and awards any appropriate punishment—dismissal, suspension or warning. Except in cases of time-keeping, the manager can only warn. The procedures established are designed to ensure a speedy resolution of any complaint, and unanimity of decisions taken, and over the years the Committee has produced a body of case law appropriate to the Company's circumstances. There are appeals procedures of this sort around, but the important question is whether the 'case law' produces the rules.

In effect, this committee is claimed to have avoided trouble over matters of discipline, but to the outsider, used to conventional arrangements for discipline, its decisions may appear to be eternal compromise. Against this, however, has to be set the very important point that if worker and management representatives are to share as equals in any decision-process, it must be expected that worker values will have some mitigating effect on what would otherwise be unilateral or paternalistic decisions of management. The fact that compromises do occur here is perhaps a sure enough indication that decisions as other than the 'right' or the 'requisite' ones, they might also have to acknowledge that rightness is something which can be judged from very different value positions, and might in any case have a high situational component (i.e. be expedient in the circumstances).

In principle, this issue is little different from the one referred to above in connection with productivity bargaining—i.e. whether workers should be allowed or encouraged to make suggestions for capital investment, or, for that matter, for product development or marketing. Once one embarks upon either exercise, different values are brought to bear and these might well lead to different decisions, with the limits falling only at the boundary of managerial discretion.

However, what management does—even when it decides unilaterally upon a course of action—will influence the profitability of the undertaking. What we may be witnessing at the moment is a recognition that future profitability is likely to be increasingly dependent upon not only worker *acceptance* of managerial authority but also worker *participation* as of right in the exercise of this authority. If this is indeed so, the manager who recognises the dependence early and seeks to vary the approach to suit the changed circumstances is likely at the end of the day to have a more viable outfit than he who bays for some unilateral moon. It is of course inconceivable that workers will *not* ask (demand) for some greater say in what happens to profits if they are asked to make a contribution to other types of decision, but in the emerging situation, there may be no profits to have a say about unless participation is increased now.

Nevertheless, in all of these managerial approaches, it is possible to detect an attempt—semi-autonomous in origin— by the administrative managers to secure acceptance of their authority by giving participation within the limits imposed by this delegated discretion: in the short run, profits are expendable in pursuit of this objective—a possible explanation of current developments which links together the rumblings of Cameron Hawley about American industry's corporate fat and Glyn and Sutcliffe's analysis of British industry's declining profitability in the 1960s. Since these managers remain dependent upon profitability in the private sector, however, this can be no more than a short-term expedient, however significant in itself; in the longer-term, some more fundamental modification of approach might be called for, and this in turn is likely to involve the directorate or top management as well as the administrators.

PARTICIPATION BY IDEOLOGICAL RECRUITMENT

The second of our three categories of participation in private concerns focuses essentially upon the recruitment of workers' attitudes to a position of acceptance of the basic beliefs implicit in the private enterprise system itself. (In this way it is distinguished from our third category, in which comparable degrees of 'participation' might form the outcome, but where the workers' attitudes do not necessarily have to change in this direction to secure such outcome.) The emphasis is therefore upon the successful application of an incentive to bring about this ideological recruitment, and this incentive has commonly taken two main forms; either (a) workers have been given some title to the economic fruits of ownership without necessarily any basis for control, or (b) they have been given some title to ownership from which they can acquire the power to share in control.

(a) *Profit-Sharing*

Private enterprise is inextricably associated with profit-making and because of this association, one of the obvious areas of decision which might be attacked or opened to participation is that which surrounds the distribution of profit. Profit-making as a goal of private enterprise endeavour may be treated as a case of displaced objectives, but the assertion that profit is the end of endeavour is still a common one. It is not, however, universally subscribed to, and there are those top managements who would argue that profit is a residual, or a standard, and not an end in itself. There are also those top managements who would regard profit as expendable in pursuit of increased incentive to worker performance or of increased loyalty to the particular enterprise. Those who act upon this belief about profit might well embrace profit-sharing as a means to these ends.

A number of firms do have profit-sharing schemes, although the total number may still be insignificant within the context of British industry as a whole. There is some suggestion that in conditions such as full employment, where loyalty has to be bought, the number of profit-sharing schemes introduced may well rise. The death rate of such schemes is however high and the total number remaining in existence over a long period of time is relatively small. The trade unions, by and large, have little time for profit-sharing, preferring to attach any share of profits due to labour to the weekly or monthly wage, and many

schemes which have been developed have therefore had associations of anti-unionism (whether strictly deserved or not). In
the face of these facts, it is perhaps surprising that profit-sharing
remains so persistent a concept in British industry. It is to be
explained presumably in ideological terms: that it is a good
thing to encourage workers to commit themselves to the capitalistic, i.e. profit-making system, to which end schemes for worker
shares in the profits are a useful means.

Profit-sharing is a term which is often given a rather wider
meaning than the words themselves might imply. First developed 150 years ago, it has come to be defined as 'an agreement by an employer with his employees that they shall receive
in partial remuneration of their labour, and in addition to their
wages, a share, fixed beforehand, in the profits realised by the
undertaking to which the profit-sharing scheme relates' (Ministry of Labour, 1920). Quite frequently, however, this type of
scheme is found in association with 'co-partnership' which implies a share in management as well as in profits, and sometimes
the two terms are used almost interchangeably.

But whilst profit-sharing clearly involves a worker participation in the profits, or the 'goodies' of industrial endeavour, it
does not necessarily imply any share in decision. The Ministry
of Labour report referred to above makes the telling point:

> It is not necessary that the employees should know all the
> details of the basis upon which the amount of their share is
> fixed: thus an employer may agree to give his employees
> one-half of all his profits in excess of a certain 'reserved
> limit', that limit being communicated only to an account-
> ant, who certifies what is due to the employees: this would
> be a case of profit-sharing (Ministry of Labour, 1920).

The word 'agree' in the above sentence could presumably read
'decide' without the sense of this being lost; all that is required
is that the proportion be fixed and known in advance, not that
it shall result from some agreement which implies a conscious
involvement of two parties in the decision. Clearly, the absence
of other information than this bare prediction as to share to be
expected, must make profit-sharing a scheme which does not
necessarily involve participation in anything other than the

goodies of the system.

(b) *Co-Partnership*

The shading of profit-sharing into some kind of co-partnership arrangements is illustrated in the case of the John Lewis Partnership Limited. In this case, too, the ideological position of the Founder, John Spedan Lewis, led to the germination of an idea, first of all of profit-sharing and later of co-partnership. On inheriting control of Peter Jones's store Spedan Lewis determined to ensure that the true profits of the enterprise went to the workers, and did this in 1919 with a capital reconstruction followed by the issue of 'share promises' to employees ranking for dividend alongside existing capital until they were converted into preference shares (provision was also made for fixed dividends and allocations to reserves before the issue of share promises).

This profit-sharing arrangement was also extended into the John Lewis business in 1929, when Spedan Lewis inherited control of that company. By an irrevocable Settlement in Trust, Spedan Lewis transferred his interests in these two stores and the Odney Estate to three Trustees, two of whom were his wife and himself, although he retained control of the enterprise through the ordinary shares of a newly-created holding company. The proceeds of the transfer were left in the business in the form of a long-term interest-free loan and the Trustees were charged with applying the profits for the benefit of employees, essentially in the manner already employed in Peter Jones.

Later, by a second settlement in Trust, in 1950, the ordinary shares in the holding company—and therefore also the control of the assets and profits—were transferred to the John Lewis Partnership Trust Limited, whose Chairman and Deputy Chairman were to be the Chairman and Deputy Chairman of the Company, the John Lewis Partnership Limited. At this time a Central Council representative of the 'partners' (i.e. the employees) was given the task of nominating five of the remaining directors to the Trust. With this Settlement, therefore, both ownership and control passed from the hands of the family owners to the Trust, which was charged with the task of using their power for the benefit of the employees of the Trust. From this time onwards, therefore, not only was the John Lewis Partnership concerned with profit-sharing, but it had also provided for a representative body to control the affairs of the enterprise.

Another example of a similar kind is provided by the Michael Jones Community Limited of Northampton. A firm of jewellers built up in nine years to a turnover of £175,000 and a profit of about £4,500 a year, by Michael Jones himself, it was turned into a 'community' in 1969. The 33 staff were made joint owners with equal votes and shared in the profits; it was even made possible for the staff to get rid of the original owner provided that 75 per cent of them agreed to this course of action. Nevertheless, a restriction was placed on the distribution of the profits: according to the community's constitution, part of the profits were to be ploughed back into the company, and equal shares of the remainder were to be given to charity and distributed amongst the staff in relation to the positions held by them in the firm.

The 'restrictions' imposed upon such 'ultimate decisions' as the distribution of surpluses from trading by the constitutions of such ventures may be sensible, prudent, and many other things—that is by the light of the business ideology whence they sprang. But they are also denials of the co-partners' rights to determine such matters in an absolute sense. Because they exist, they can be, and are, interpreted as a reflection of the underlying paternalism from whence comes the 'give away' of the firm's assets and (with certain limitations) the power to determine corporate policy.

From the workers' point of view, the acid test of such schemes could well be whether the future of the enterprise can be determined independently of their wishes. Certainly, there are examples on record of such enterprises foundering upon the rocks of economic competition and being closed down in spite of the original owner's insistence that his workers were his partners and that 'their rights, their interest and welfare were his constant concern' (Pafford, 1970, p. 22). The example from which this quotation is taken is the Quaker-owned silk spinning firm of Ford, Ayrton and Company of Low Bentham, near Lancaster, which at the time of closure in 1970 had experienced

25 years of co-partnership and profit-sharing under the guidance of Charles Ford, a son of the Founder and Chairman and Managing Director until his death in 1964. In 1917 the Company had started to pay a bonus from profits, and this was extended and made more permanent two years later when, also, workers were invited to become shareholders and three were appointed worker directors (two by election by their fellows and one by election by the permanent directors). These directors attended all meetings of the Board, but were not allowed to interfere in the day-to-day management of the mill. Nor were they, in the face of increased competition from foreign silk spinners, able to prevent the closure of the mill.

A similar, if less cataclysmic, fate appears to have overtaken the Kalamazoo experiment in co-partnership after 24 years. Kalamazoo was founded by two Quaker families, the Morlands and the Impeys. In 1948 the Kalamazoo Workers' Alliance was established by F. L. Impey, and on this body was settled 32 per cent of the Company's stock. In the intervening period, the dividends received on this holding have been applied as to 80 per cent in bonuses to members and 20 per cent in purchase of more stock. It was envisaged that this would create an accumulating trust under which the Alliance could go on accumulating stock 'for ever'. However, the death of the founder revealed that this assumption could not hold. A note appended to the 1972 results, states: 'As a result of the death of the Settlor, Mr. F. L. Impey, in September, 1971, the Kalamazoo Workers' Alliance will not be able to accumulate income in 1972–3 and later years for the purchase of more Ordinary Stock in the Company. The present equity holding of 51.5 per cent is therefore unlikely to increase.' By implication, any well-intentioned creation of a trust arrangement by an owner is to be regarded as good for his life-time. Beyond this, there could be reversion as in the case of Kalamazoo.

(c) *Worker Influence on Directives*

This focuses attention on the nub of the whole problem in private enterprise concerns. It is to be found in the last Labour Government's consultative document as in the earlier Labour Party report on Industrial Democracy (1967). The problem

arises from the fact that the belief in the correctness of allowing
an owner to do what he will with his own has pervaded our
whole approach to the structuring of the industrial system.
Reflecting this belief, the law as it touches upon the structure
of the industrial enterprises of Britain has sought progressively
to protect the interests of the owner and shareholder not merely
against workers but mainly against those who seek to use the
capital of the shareholder to develop productive enterprises, the
company promoters and their descendants, the company
directors.

In law, the members of the Company are the shareholders—
and only the shareholders—and the charge upon the directors
is to serve their interests to the best of their ability as fiduciaries
and agents. Whilst other interests than those of the shareholders
may be taken into account in directorial decisions, the 'ultimate'
test is whether or not the members' interests are best served by
taking them into account in decisions. Ultimate authority is
therefore firmly with the shareholders, whether they choose to
exercise it or not, and the Board of Directors may exercise it in
the interval between general meetings, as the agents and trustees
of the shareholders.

This upholding of a fundamental value in law is a fact of
considerable significance in relation to the problem of 'representing the worker interest' in the basic decisions taken within
private enterprise concerns. The law, per se, does not disbar
workers from membership either of the Company (they can
become shareholders in the normal way or by some special
arrangement involving the purchase or donation of shares) or
of the Board of Directors (although they would be subject to
appointment or confirmation in office by the annual general
meeting of the shareholders). The Company's Articles of Association, which are privately drawn up albeit in such a way as to
meet the requirements of the law, may, however, make it impossible for workers to be co-opted, appointed or confirmed in
office as Directors. Also, the legal definition of the director's
fiduciary responsibility to the shareholders which requires him
to act consistently in the best interests of the Company as a
whole, would render it impossible for the workers' representatives to act, as Directors, in the interests of 'the workers' in any

circumstance where this interest might be held to be inimical with the interests of the shareholders. Where such a conflict might obtain, the workers' representative on the Board of Directors could have but one duty and one course of action— to serve the shareholder interest. If, therefore, the law (and where appropriate the Articles of Association) is not changed, there is only one opportunity open to the enterprise which wishes to secure worker participation in the 'ultimate decisions' affecting the enterprise: this is the device of making all employees shareholders. But to do this, it would be necessary to secure the support and acceptance of the existing owner-shareholders, and this is likely to prove more difficult to bring about than the securing of the support and acceptance of the Board of Directors. For this reason, few companies have in fact made their employees shareholders with the intention of modifying the decision structure at the corporate level.

Where this has been done, it has usually involved the private limited liability company—the enterprise which has a limited number of shareholders. It has been done, and a not untypical example of this form of development (and the one best reported) is the Scott-Bader Commonwealth. In this case, as in many of the others, the motivation for the change is a strong personality at the top who develops as his ideology, the belief that private profit-making is either wrong or too restricting. For this position, there is an ancestry going back at least to Robert Owen, but the present day ventures of this sort do normally involve a strongly religiously-motivated corporate management, often provided by a family, who—in the words of Ernest Bader 'divest themselves gladly and willingly' of their absolute ownership of the enterprise in order to share this and the power of decision with some wider group (such as former owners plus workers). In a larger undertaking with a much more dispersed shareholding and a larger number of shareholders, or with a strong institutional shareholding securing the necessary agreement to divest might prove much more difficult than in the small private company, such as Scott-Bader or Farmer and Sons Ltd.

In the Scott-Bader case, 90 per cent of the shares held by the family in the original company were divested in this way to the 'members' (owners, managers and workers) of the newly created 'Commonwealth' which would henceforth run the enterprise. Membership of the Commonwealth is open to all over 21 years, who are willing to give 8 hours voluntarily to the commonwealth each year and a shilling in the event of it being wound up, but they do not thereby acquire any continuing financial rights in the Commonwealth. The new Board of Directors comprises two members directors elected triennially, three Founder (i.e. family) directors, and two family-nominated directors which runs the enterprise through a number of appointed and elected committees. The most important of these is the General Council, comprising eight management members and eight ordinary members elected in constituencies, which discusses all matters to do with the firm's and its employees' welfare, including financial matters and matters of discipline.

The profits made are distributed according to the wishes of the members in the following way:

60 per cent for taxes and Company's reserves;
20 per cent to employees on the basis of one-third to everyone and the rest in proportion to pay received;
20 per cent to Charities, selected in the annual general meeting, and justified constitutionally on the grounds that 'a socially responsible business cannot exist solely for its own interest'.

Given the main objectives of using this device of the Commonwealth as a means of improving the man, it is difficult to determine whether it is 'successful'. By usual commercial yardsticks, it is doing well enough, wages are high enough to secure labour, people working there seem satisfied enough. But it developed out of one man's idealism, and the ultimate test must therefore be whether it approaches realising this end, and this must remain pretty impossible to measure in any simple way.

What it does illustrate, however, is that top management can produce modifications both of the structure of ultimate decision and of the treatment of surpluses, residues or profits even within the present framework of legislation. The main question which other managers seek to have answered by such organisations is what its future potential or capacity is, and it is this question which it is most difficult to answer because of the idealism which shapes the kind of future aimed at. Consequently, the question is similar to that which might ask for an objective answer as to whether oranges or apples taste better.

What all of these kinds of experiment also tend to imply is, firstly, that there is a hierarchy of decision and action in which the former must precede the latter, and secondly, that top management must retain the rights of decision, no matter how these are constrained by influences from other influence groups. Given the particular form of conceptualisation of enterprise which Barnard launched upon industrialists in 1938 (and prose-letised by disciples since then) the top management of an enter-prise *can* find comfort in the necessary implication that someone must reconcile the varying interests of shareholder, worker, customer, community, supplier, etc. and carry out, in fact, what has become known as the 'statesman role' of top manage-ment.

This is usually supported tacitly but effectively by statements about top management's moral or social responsibilities. Most of these now diminish the importance of the shareholders as the source of top management authority. As Levy has commented, for example,

> The adherents of the doctrine that directors are fiduciaries of the corporation itself sometimes broaden this view to mean that directors must look not only to the interests of shareholders but also to those of the employees and workers of the corporation, as well as to those of consumers, and generally of all third persons who come into relation with the company, and finally to those of the community (Levy, 1950).

Such issues have to be considered in terms of morality, rather than in the more precise requirements of the law. In spite of the legal restrictions, it is not necessary as Copeman has said, 'to accept a particular view of directors' legal responsi-bilities in order to believe that they have wide moral responsi-bilities'. Nevertheless, moral imperatives can be detected, as in the American Management Association monthly bulletin com-ment in 1951 on this question:

Whom should directors feel they represent—the stock-holders, the management, the general public, the particular community in which plants are located, the employees, the suppliers, the creditors, or the customers? The answer is that directors have responsibilities towards each of these groups and many times the boards of large corporations must weigh the interests of one of these groups against the other.

This is to state something of the pressures upon top management and to indicate that in accommodating them, they must perforce give attention to pressures other than those for high profits. Unlike the discussions of this issue within the context of worker ideology, however, this does not raise the question of rights: it is merely to establish the presence of interests and to underline the moral position adopted by top management in exercising their high levels of discretion. Whilst this may mitigate the severity of application of 'pure profit' motives associated with private capitalism, it does not go as far as many worker demands would suggest. In fact, many of the ideologies expressed about the top management role could be conceived, either as individual views (subject to removal with the individual from the position) or as temporary philosophies adopted by the top management in line with the times and circumstances (but equally capable of change).

This could be true for example of the views expressed in the 1951 annual statement of the chairman of Tube Investments Ltd., by Mr. I. A. R. Stedeford: 'The purpose of an industrial company, as I believe most of us see it, is to produce more and better goods at relatively lower prices; to strive to provide a good and secure living for its employees; and to provide a safe investment with a reasonable return for those who found the money for the enterprise' (12 December, 1951). Stedeford's viewpoint was supported by F. A. Hurst, chairman and managing director of Samuel Osborn & Co. Ltd., the Sheffield steel firm. He pointed out that: 'It is only by having in mind the facts Mr. Stedeford puts forward that there is any worthwhile return on capital' (*Financial Times*, 20 December 1951). In similar fashion the American firm of Johnson and Johnson, with international (including British) subsidiary activities, has adopted a 'Credo', in which the company's first responsibility is recognised as being to its customers, its second to its work-people, the third to management, the fourth to communities in which the firm lives, and the fifth and last responsibility is to the stockholders.

Action to translate these views into constitutions are relatively few. Early in 1951, the Glacier Metal Co. Ltd. issued a 'Policy Document' to its employees, in which the company is described as 'seeking to establish an increasingly democratic government of the company community, which will award fair responsi-bility, rights and opportunities to all its members, consumers and shareholders.'

Other examples are provided by the more fundamentalist experiments referred to above—e.g. the Zeiss Foundation of Jena (Goyder, 1951), the Scott-Bader Commonwealth (Blum, 1968), Farmer and Sons Limited (Company Record, 1962) and Factories for Peace (Rowen Engineering Company, 1965). There are also many milder variations on the same theme which seek something similar through the development of 'joint consultation' which may or may not be as carefully codified as in the Glacier Metal Company (Jaques, 1951).

But the fact should not be lost sight of, that, if these de-velopments reflect a top management accommodation of cur-rent pressures from different sources, then a change in the relative force of these pressures may well lead to a different accommodation in which the worker interest is diminished in strength. Since these ideologies do often have individual asso-ciations, they are as yet not universally accepted even by top managements, and certainly not by the shareholders or those who act as their watchdogs. Thus, we find the *Investors'* *Chronicle* criticising the Glacier Company Policy document from the shareholders' point of view in its issue of 7 July 1951, but at the annual meeting on 11 July, although shareholders were critical of the company's conservative dividend policy, none mentioned directly the 'Policy Document'. The reaction to Stedeford's statement was similar. It was challenged by various shareholders in the *Financial Times* as well as by 'Lex' in his column, the gist of the long debate being that the shareholder interest should be put first. Thus, the Credo of the company director may not differ from that of the shareholder, but when they do, it could well be the beliefs of the director which change before those of the latter.

The final stumbling block to increased participation at this directive level is therefore likely to be that shareholder who is in the company for the return on capital and not the predisposition or predilection of the Director, who, possibly because of his closer proximity to the problems of running a business on the basis of at least a minimum of consent, might be more ready to contemplate such modifications. Since there is some evidence to suggest that participation in ownership, through dispersed shareholding, can strengthen the desire for financial return, action to increase the level of participation in this may, para-doxically, raise higher barriers to participation in government or direction. Since dispersed shareholding also seems to ac-company increased directorial autonomy—of the Burnham 'managerial revolution' type—the immediate future may well lie with a development of the top manager's 'statesman' role, through which the various interests of the various categories are autocratically 'taken into account': equally paradoxically per-haps, this development would allow the 'give away' pattern to develop, provided always that the shareholders could be got to agree. The experiments which have been noted above do rather suggest that a benevolent paternalism is more likely to associate with relatively small boards of executive directors than with large bodies of shareholders, although the very largeness of the latter may in turn contribute to the opportunities for the former.

THE TRADE UNION CHALLENGE

In the face of such guarded attempts to spread decisional auth-ority without complete abdication of rights deriving from ownership, it might seem reasonable that the main challenge to managerial authority would be that which springs from the trade unions and their activities. Many trade union constitu-tions declare these organisations to be concerned with the destruction or overthrow of the traditional capitalist system, usually by 'bringing the means of production, distribution and exchange into public ownership'. Whilst it is true that a number of other objectives are also indicated, there is nevertheless a *prima facie* case for asserting that the trade union is concerned to challenge the whole authority structure of the private enterprise system.

Nor should the influence of the trade unions in bringing about such changes in the distribution of power be overly minimised. But in this present context, it is sensible to distinguish the industrial and the political activities of trade unions from this point of view. The organisation of workers into fairly coherent political groupings, supporting, for the most part, the Labour Party, has, from time to time, helped to sustain a Labour Government in power and thus permitted it to change some of the ground rules applied to the distribution of power in industry. The marshalling of these same workers behind a policy-modifying objective within the industrial context itself, whilst it may have increased the 'order' visible in industrial relations, has had less dramatic impact upon the distribution of power and authority in the private enterprise system.

As a result of political action, through governments, some of the means of production distribution and exchange have been brought into public ownership and some important legal curbs have been placed upon the exercise of authority by companies and their directors. As a result of industrial action, on the other hand, there has grown up an elaborate procedure through which trade unions exercise a right to discuss certain matters affecting their working lives with their employers but these form only a rather limited set of substantive decisions which do, in fact, affect the working conditions of their members. At the present time, and in so far as these conditions are universally restricted, they are constrained more often by the consequences of legislation than by the consequences of industrial action. The influence of the labour movement on political decisions should not be discounted, but from the standpoint of our present concern, it is peripheral to the main issue of challenge to authority within private enterprise.

The main form taken by the workers' challenge in industry itself to managerial authority has been that of concerted withdrawal of labour within the wider framework of voluntary collective bargaining. Such withdrawals have been designed to influence the employer to change a decision or proposed decision in favour of a point of view put forward by the employees as a whole or by some class or category of them. Rarely has this challenge been extended beyond the confines of the single industry—although this did happen with the Triple Alliance and the General Strike of 1926—and it has often been confined to workers within a single Company or plant.

Voluntary collective bargaining does not in itself represent a direct challenge to the exercise of managerial authority, but is rather a mechanism for containing the challenge or for giving greater order to the process of bringing worker influence to bear upon managerial decisions. In this sense, it could be interpreted as a process of 'buying off' the worker challenge to that authority by laying down procedural rules which both parties agree to abide by in working out compromise decisions about substantive terms and conditions of employment. Once the trade union or the workers accept this system they cease to pursue a revolutionary course of action in relation to authority, and agree instead to limit their challenge to such issues as can be tackled within the framework of procedural rules or by recourse to strike action when the matter cannot be resolved within that framework.

In greater particular, this voluntary collective bargaining system has placed two main limitations on the worker challenge.

Firstly, it has tended to confine the influence to a limited range of issues and correspondingly left to unilateral management decision all other issues. The matters which have been negotiated over have included wages, hours, holidays, and certain aspects of manning (e.g. apprenticeship, recruitment, promotion and advancement). The matters which have been left free from direct worker influence have included such worker-related matters as production scheduling, safety, health, and pensions, as well as dismissal and redundancy questions for the most part. To note that the frontier between these two categories has moved over time, is not to deny the point in principle. In comparison with practice in the USA and many continental European countries, the issues left for unilateral employer determination in Britain are extensive (Kahn Freund,

1972). On such comparison, the effect of voluntary collective bargaining arrangements in Britain has been to restrict the breadth of the worker challenge to the employer's authority or prerogatives.

Secondly, it has also tended to restrict such challenges as have been made to matters which are chiefly of administrative management concern and only indirectly of directorial concern. The list of negotiable issues is confined largely to matters which have more to do with the means, modes and methods of achieving the ends of the enterprise than with the objectives themselves. Issues such as the closure or relocation of a production unit have—until recently—attracted little direct challenge from the trade unions who have confined their negotiations on these issues to procedures for redundancy or transfer or compensation. In the nature of the situation, closures fall outside the rules and procedures and can only be attached by mobilising opinion by protest marches, demonstrations, and the like. A fairly fundamental acceptance by the workers in their trade unions of the principle that an owner can do what he will with his own— provided he meets certain norms of conduct—has meant that legislation has often been applied to the regulation of these matters in the face of trade union opposition or indifference.

Taking these two together, it might be argued that British trade unions have developed a facility—not to say an obsession —for the development of procedures through which matters of substance might be decided on a 'joint' basis, but have at the same time put up a rather poor showing in achieving the substance in any abundance or in any range. The current pursuit of money wage increases through the rump procedures which the 1960s have left them with is no more than a latter-day illustration of this displacement of what most union constitutions and rule books declare to be their objectives as voluntary associations.

Although recent years have witnessed something of an upturn in the amount of strike activity—just how much of an upturn it is impossible to say given the manner in which such data as we have available is collected and assembled—the relatively low strike record in British industry since 1926 provides further support for the contention that British trade unions have offered little real challenge to the employer in the past four decades. The challenge is mainly one related to procedures and not particularly to substance.

What the trade unions have sought to do, essentially, is maintain the real standards of living of their members. This has been sought within the framework of collective bargaining resting upon the ultimate sanctions of strike or lock-out. In the long term, there is little doubt that the real standards enjoyed by British workers have improved; but in the same period there is little evidence to suggest that this improvement has been achieved at the expense of the interest of the shareholder or profit-taker. One possible inference to be drawn from this is, therefore, that the present institutional arrangements do not so much represent a challenge to authority in the private enterprise system as a collusion or co-operation to exclude a further interest, e.g. that of the consumer. Bearing in mind the dual role of the worker, as both employee and as consumer, and reminiscent of the guild socialist's recognition of the need for a partnership between state and guild to secure the worker interest, it could well be that only by accepting the Governmental role as closer of the open end to the negotiation will the trade unions have a long-term hope of turning collective bargaining into an effective challenge to owner-shareholder-director authority.

The trade unions have also been concerned to 'protect' their members, or to enhance the security they possessed as income earners. Industrial action has been directed towards controlling manning (e.g. through apprenticeship control) and ordering dismissals (e.g. by establishing procedures governing redundancy). It is, however, significant that a few fairly simple pieces of legislation in the 1960s—the Contracts of Employment Act, the Redundancy Payments Act, and the dismissals part of the Industrial Relations Act—achieved much more 'at the stroke of the legislator's pen' than the trade unions had been able to achieve in most cases of negotiation on this issue. It is also significant that British trade unions have given more attention to the control of labour intake than to the control of dismissals: whilst this reflects the peculiar history of the deformation of generalist occupations and the introduction of 'illegal men', it also reflects the ethos of expansion in which British trade unions developed and points to the current difficulties of accommodating a situation in which Britain is no longer the workshop of the world located at the centre of an economically-subject empire which provided cheap food and raw materials and a ready market for products. By the post-war period, the approach of the trade unions had left British workers relatively 'exposed' because managerial prerogatives in this area remained largely intact.

The clearly visible change in climate in industrial relations in the 1960s and early 1970s, which was almost independent of the political complexion of the Government of the day, could be linked back to this. This period witnessed a general increase in what was usually referred to as unofficial or even irresponsible action on the part of workers. This in itself, exemplified in more numerous short, unconstitutional or unofficial strikes, represented a change, a move away from the ordered national and district bargaining processes. It produced its own response from management. This might be said to have ranged from more local bargaining but conceived on the traditional (national bargaining) lines to productivity bargaining in which new conceptions of authority-sharing were sometimes to be discerned. The Government patently failed to close the open end of the bargain, however, and an out-of-control, inflationary situation developed as workers ran like mad to stay where they were.

Even if some of these developments appeared to change the distribution of authority within administrative levels of decision —either because managements sought to 'involve' workers more or more simply because managements saw their power to order the situation swept away by unofficial action, other developments on the industrial relations front during the late 1960s and early 1970s promised a subtle but significant change in the direction of trade union militant action. This focused upon the techniques of 'the work-in' in its milder form, and 'the sit-in' in its more significant mien. Particularly in the latter case, the workers discovered the value of using the 'capital' as the bargaining counter in considering proposals to close plants.

The sit-in is not, of course, new in itself; it has been employed many times and for many purposes. In the 1970-2 period, for example, it was used extensively, particularly in the Manchester area, in association with the AUEW attempt to secure local agreements in the engineering industry after the termination of the national procedural agreement in that industry. In the same period, the method was also used to pressure employers to grant pay increases which had been demanded. But the more significant development in this area was the use of the sit-in to back a demand for continued employment: 'we refuse to allow the Company to put us on to the labour market' was how one Midlands steward summarised the objective. The work-in has then usually represented a variation of this method, and is, of course, best exemplified in the Upper Clyde Shipbuilders experience (which is reported elsewhere). Amongst the early examples of this method were the Plessey (Alexandria), BSA (Birmingham), BSC's River Don (Sheffield), Allis-Chalmers (Mold) and Fisher-Bendix (Liverpool) experiences. Later examples were provided by Ransome, Hoffman and Pollard (Neward), Tube Investments (Wolverhampton and Walsall), BL (Thorneycroft) and Briant Colour Printing.

The simple demand to vary the terms of a board-room decision and the use of the sit-in to support it is illustrated in the confrontation between British Leyland and its trade unions over the proposed merger of Transport Equipment (Thorneycroft) with the American Eaton Corporation of Cleveland. Under the merger terms, the Eaton Company was to continue to employ about 700 of the workers who were engaged in the heavy transmissions side of the Basingstoke plant and to supply transmission components to BL. Certain other fringe activities were to be transferred to other BL factories with the loss of 350 of the 1,100 jobs in Basingstoke, but the BL Company offered generous terms for voluntary termination of employment by those affected.

The official leadership of the unions involved, led by Mr. Reg Birch, declared that they would not tolerate this deal and would support whatever efforts the Thorneycroft workers made to thwart it. Such readiness on the part of the AUEW to support the members could reflect the experience which the union had had with the company between 1962 and 1968, when the Eaton Yale and Towne Company, having taken over the ENV plant in Willesden, closed it and transferred the work to other plants, after six years of disputes mainly about work practices. The powerful BL shop stewards' executive committee which secured official backing complained of lack of consultation by management about financial policy generally and demanded that more information be given to the unions about the future of employees. Lord Stokes declared that this merger was necessary because BL could not afford to develop this activity and remain competitive, and that it was the need for commercial confidentiality which prevented the Company from giving advance warning to the unions, although they were the first to be told about it when the need for confidentiality was passed.

The Joint Chairman of the Committee, Mr. Eddie McGarry, declared that if they got the 'brush-off' from Lord Stokes, he proposed to recommend to the shop steward convenors in all BL plants that 'something be done in order to bring home to him the importance of taking us into his confidence'. What Mr. McGarry seemed to have in mind was the withdrawal of co-operation in the matter of developing new and modern wage structures in the BL plants. At a meeting of the stewards on 9th August, it was agreed to mass picket the plant on the following Monday, hold a one-day strike throughout BL on 28 August, and give full support to the Jaguar plant stewards whose plants had by then already been on strike for seven weeks.

On 14 August 1972, immediately following the two-week holiday shut-down, the workers accepted the suggestion of the AUEW divisional officer's recommendation of a sit-in, and occupied the boardroom, offices and factory. The initial intention was to persuade the BL Company to unscramble the deal made with the Eaton Company, and the eventual intention to secure improved conditions of transfer from the BL Company's employment. In the first month, about 300 workers drifted away from the plant, although not all of them went to other jobs, leaving somewhat in excess of 600 in occupation. The Eaton Company early agreed to provide employment for about 700 of the BL employees, but (no doubt reflecting the experience with the Willesden plant) the union stood out for a guarantee of 1,000 jobs, improved conditions of transfer, and stronger guarantees of continued orders from BL to the new company.

The sit-in lasted for ten-and-a-half weeks, the workers agreeing by 400 votes to 200 on 27 October to accept the latest offer from BL and end the sit-in on the following Monday. In the settlement, the union dropped its objections to the take-over by Eaton Corporation on 15 January, and British Leyland increased to £200 its offer of special payments for disturbance to the 738 workers to be taken on by Eatons and a few others still facing redundancy and agreed to guarantee orders to secure those 738 jobs for three years. The BL Company has also agreed to meet the Unions at any time should the orders to Eatons for components fail to secure the jobs provided by Eatons. The increased cash offer contains an element of hidden agenda: on the one hand it will help to alleviate the hardship caused to workers during the period of the sit-in, and on the other it has been vaguely tied to an undertaking from the workers that production will be brought back to normal within six weeks of a resumption of normal working. It was also expected that the settlement would reduce the likelihood of sympathetic action from other workers in the BL combine, particularly at Southall where such action had been planned, and of lay-offs consequent upon the drying-up of the supply of components.

Although in the event this rather lengthy sit-in did not achieve its original declared objectives of maintaining work for 1,000 employees and preventing the £5m. sale of the plant to Eaton's, it did result in some modification of the terms of the fundamental decision about the plant's future taken, unilaterally, at the corporate level in two companies.

The importance of what happened in these cases is therefore in the kind of demand the action is designed to support. This was and still is a demand for the continuation of employment, even when the directors determine that the economics of the situation demands that the plant be cut back or closed in line with the right of the owners to dispose their property in the manner best suited to their interests. As it was put by a shop steward at the Tube Investments, Walsall, plant, the strategy is to 'operate if we can, and occupy if we can't'. The demand has been voiced often enough by trade unionists in the past, but rarely has it been supported so universally and so firmly as in this period.

The use of this method of challenge did not go unremarked by the committed private enterprise supporters. For them there was a novel principle to be discerned which challenged the basic tenets of the private enterprise beliefs. For example, in the middle of June 1972, 150 employees of the Briant Colour Printing Company in South London began a work-in after the Company announced that it was going into voluntary liquidation. The unions involved gave the work-in official backing. In the following month, the workers involved in the dispute began picketing the plant of Robert Horne, the Briant company's main creditor, whom the workers blamed for forcing the Briant company into liquidation. The effect of their action was to prevent lorries delivering or collecting paper from the Robert Horne plant, whose Chairman commented: 'The whole business community should take note of the serious implications of this novel case of blacking, where workers in a bankrupt company refuse to accept the liquidation and take damaging action against their biggest creditor, without any apparent motive beyond trying to keep them going.' During this work-in the Briant workers produced leaflets in support of the dockers jailed by the NIRC and achieved further publicity thereby. They were, however, unwilling to withdraw the Robert Horne pickets on the latter's chairman giving written guarantees of help to find a buyer for the Briant plant, and the sit-in continued for at least fifteen weeks until the end of the year. It ended with a proposal to re-launch the company under new ownership, but with a management committee composed of representatives of three printing unions and managers put in by the new owner, reporting to the Board of Directors.

By 1972 there was increasing evidence that this method and this objective were beginning to have more attention from the official trade union bodies after a period of vague acceptance of redundancies and closures. Thus, the Scottish TUC in April, 1972, was faced with the choice between marching towards 'socialism in our time' by means of such direct action as this, or working within the present economic system for the benefit of trade union members; a vote on the issue was avoided only by a procedural device. The AUEW national conference in Hastings in June, 1972, agreed unanimously that where a factory was closed by merger, all members in all plants of the combine would refuse to do transferred work. The Nottingham-shire Mineworkers in the same month pledged themselves to fight redundancies and pit closures which were implemented for reasons other than exhaustion of reserves. Thus, from being the concern of relatively small, often localised, pockets of militant stewards and workers, the idea of supporting the 'right to continued employment' by militant action, including sit-ins and work-ins began to spread through the official bodies of the trade unions.

It is possible to argue that this is all recent experience and therefore difficult to assess in its long-term implications. It can also be argued that the underlying condition of high levels of unemployment is a sufficient explanation of its occurrence; by implication a fall back in the unemployment rate might be expected to bring the end of sit-ins. But whilst all this might be true, there is the possible argument that the seizure of capital assets for use in bargaining over a termination decision is a tactic which has now been learned and will continue to be applied whenever these kinds of 'ultimate' decisions are made. This might be put in terms of its helping *de facto* to establish a *collective* quasi-property right in continued employment for the workers in extension from such rights created for the *individual* by the Redundancy Payments Act of 1965. In this vein the method does counterpose a new foundation for ultimate deci-sion to set against the real property rights of the owners. In this sense it is not revolutionary—as so many have sought to argue —but merely an extension beyond the traditional level of penetration to challenge management decisions.

CONCLUSION

In conclusion, it may be suggested that much more experiment has been made with sharing in decision at the administrative/ executive levels of decision-taking in British industry than at the directorial level. Whether the ideological orientation at this level has been associated with joint consultation within a private enterprise system or with joint regulation (in Flanders' term) at the plant level, there has been a significant amount of exploration of ways and means of bringing the workers in on the decision act. Even if this has been motivated by a managerial recognition of the 'difficulty' of meeting objectives and of implementing policies in any other way, thus often giving the appearance of a sort of paternalism, it has occurred and has been 'accepted'.

At the higher decision level, on the other hand, relatively little has happened. It is true that there are a few experiments which owe something to doctrinal influence (whether religious or political) but the outstanding characteristic of the private enterprise system is the persistence of not only the primacy but also the sanctity of the ownership principle as a foundation for the exercise of ultimate authority. There are really very few experiments at this level which involve passing over some authority to the workers on the basis of 'right', and few also which have entailed trade unions in pursuit of an 'interest' in this area. In other words, none of the main competing ideologies in this field have really made much impact upon the distribution of authority to determine ends in organisations of the private enterprise type.

Furthermore, it is difficult to see in the few examples of change by either self-will on the part of the owners or successful challenge on the part of organised labour, any very clear trend in this direction of increased participation in overall decisions. The co-partnership and commonwealth ventures reveal themselves as heavily dependent upon quite strongly personalised credos and as little absorbed into the mainstream of enterprise philosophy. The main trade union and worker challenges are directed towards changing administrative decisions for the most part and where they are not they could be closely identified with a particular kind of macro-economic climate, likely to run into the sand with the return of fuller employment conditions. It therefore remains to be seen whether on the one hand there is any very strong motivation in the society to change the position by a revision of companies legislation or on the other there is any great willingness to accept Continental influences in the direction of co-determination in the new concept of the 'European Company'.

REFERENCES

1. C. Barnard, *The Functions of the Executive*, University Press, Harvard, 1938; Fred H. Blum, *Work and Community*, Routledge and Kegan Paul, 1968; W. Brown, *Some Problems of a Factory*, I.P.M., 1953; W. Brown, *Exploration in Management*, Heinemann, 1960; J. Burnham, *The Managerial Revolution*, Penguin, 1939; N. W. Chamberlain, *The Union Challenge to Management Control*, Harper, 1948.

2. G.K.N. Shotton, *A Charter for Work People*, G.K.N. Shotton, Ltd., 1969; George Copeman, *Leaders of British Industry*, Gee, 1955.

3. Allan Flanders, Ruth Pomeranz and Joan Woodward, *Experiment in Industrial Democracy*, Faber and Faber, London, 1968.

4. Alan Fox, *The Milton Plan*, Institute of Personnel Management, London, 1965; J. French, J. Israel and D. Aas, 'An Experiment in Participation in a Norwegian Factory' *Human Relations*, 13, 1960, pp. 3–16.

5. Andrew Glyn and Bob Sutcliffe, *British Capitalism, Workers and the Profits Squeeze*, Penguin Books, Harmondsworth, 1972.

6. G. Goyder, *The Responsible Company*, University Press, Oxford 1961; Chas. L. Hughes, *Goal Setting*, American Management Association, 1965.

7. E. Jaques, *Measurement of Responsibility*, Tavistock, London, 1956; E. Jaques, *The Changing Culture of a Factory*, Tavistock, 1951.

8. Otto Kahn Freund, *Labour and the Law*, Stevens, London, 1972.

9. John Spedan Lewis, *Partnership for All*, John Lewis Partnership, London, 1948.

10. Ministry of Labour, *Report on Profit-Sharing and Labour Co-Partnership, in the U.K.*, Ministry of Labour, 1920, Cmnd. 544; A. B. Levy, *Private Corporations and their Control*, Routledge, 1950.

11. — *New Life in Industry*, Scott-Bader Commonwealth Ltd., 1955.

12. J. H. P. Pafford, 'Ford, Ayrton and Co., Ltd.' in *Co-*

partnership, 539, London, January 1970, p. 22; N. Parkinson, *Ownership of Industry*, Eyre and Spottiswoode, 1951.

13. T. T. Paterson, *Job Evaluation* (2 vols.), Business Books, London, 1972.

14. S. Richbell, 'A Study of Implementation Problems in a Productivity Bargain', University of Wales Ph.D. thesis, Cardiff, 1971.

15. G. F. Thomason, *Experiments in Participation*, Institute of Personnel Management, London, 1971; Kenneth F. Walker, *Industrial Democracy, Fantasy, Fiction or Fact*, The Times Newspapers, Ltd., 1970.

Workers' Participation in Western Europe

CAMPBELL BALFOUR

Workers' councils in Western Europe have a long history and have been developed much further than is the case in Britain. There are many reasons for this, stemming from history, culture and war. As we shall see, the new flowering of democracy after 1945 in the occupied countries led to the revival of workers' councils in industry as part of the democratisation of everyday life.

In contrast, it may be argued that industrial democracy is not possible in a totalitarian system; when workers have the right to challenge and influence managerial decisions, such challenges may adversely affect the national plan. This point is made in the classic study 'Workers' Councils' by Professor Sturmthal who writes, 'fair chances of success for either side, absence of police interference, and the refusal to use governmental power for the benefit of one side—these are characteristics of self-government which a dictatorship is unlikely to tolerate'.[1]

The success of workers' councils in Western Europe is still a matter for debate. Some employers' organisations claim that the right to manage and take decisions is inherent in a free enterprise system, although other employers agree with Reginald Maudling, ex-Chancellor and Home Secretary, who stated in late 1972 that the classic market economy was no longer viable, and the market had to function in a framework of Government supervision. This statement is characteristic of the mixed economies found in Western Europe, where governments are often coalitions of conservatives and social democrats, where employment and economic growth are at a higher level than in Britain, and where governments intervene in economic planning and government aid even (or especially) in Gaullist France and Adenauer's Germany.

Though the debate continues over participation in industry, successive government committees in different countries have recommended an extension of the powers of workers councils. The Treaty of Rome, now signed by nine countries, encourages the principle of representation, while two of the countries who have stayed outside the Common Market, Norway and Sweden, have long and well-established systems of industrial democracy. In the particular context of workers' participation, Britain stands almost alone in Western Europe for its lack of institutions or general support for the idea. Changes are certain to come when employers and trade unions of the various countries meet more frequently and exchange ideas and expectations. The spread of multinational companies, already a form of market without frontiers, will help to speed this process.

We do not deal with developments in Yugoslavia, though some writers see this as the answer to the employer/worker dichotomy. While there appears to be a greater share for workers in decision-making, it is evident that numbers of councils succeed in voting themselves higher wages than is justified by the rate of production. Consequently, while morale may be high, inflation is running at the damaging rate of 15 per cent. Yet some of the lessons of the Yugoslav experiment might be tried in Western Europe, and especially in Britain. Sturmthal points out that members of works councils in Yugoslavia 'learn by doing' and are able to confront real problems in their council work. He sees this function of the council as a 'powerful educational device'.

The economies of Eastern Europe are aware of the possible inflationary effects of workers' councils and seek to contain and direct this possibility by education in management problems, e.g. 'If we wish to change the workers' mentality we must develop their understanding of economic problems. They must realise that from now on it is essential that the undertaking should pay its way. . .'[2]

We can say that employers tend to be more progressive in terms of 'social responsibility' in countries such as Belgium, Holland, Germany, Denmark and Norway, and more autocratic or anti-union in countries such as France and Italy. There are historical and cultural reasons for such differences, some of which can be explained by the fact that the two latter countries have huge communist parties, large agricultural labour forces, politically-motivated trade unions, along with important Catholic groups often influencing governments and channelling christian worker opposition to communist-controlled unions.

There is the influence of the structure of industry. The average size firm in France, for example, tends to be small and the employer has the attitude of the small 'go-it-alone' businessman who was often the backbone of the Poujadiste anti-government, anti-labour group. The small family business in France, from tax as well as other reasons, does not react well to the suggestion that they should share financial information at works council meetings.

While the powerful French Communist Federation of Labour (CGT) has historically been hostile to co-operation with employers and has preached the class struggle and the control of production by the workers, there have been signs of change in recent years. There are several reasons for the change in communist attitudes to works councils: the growing coexistence between Russia and the USA, and the thawing of the cold war in Western Europe, culminating in West Germany's Ostpolitik; the majority support given by workers to the growth and development of the Common Market, so that the communists have changed from opposition to qualified support and accepting seats on EEC committees; lastly, the shift in Marxist ideology from 'exploitation' and resistance to employer demands to 'alienation' or improving the workers' life at work. The 'alienated' worker, in the absence of the state controlling industry, can negotiate for a better life in the factory which can now include 'participation in decision-making' in works councils. According to L. G. de Bellecombe, this shift was seen clearly after the riots of 1968, when France was near to anarchy, but stemmed from the Grenoble conference in 1966.[3] He quotes a rapporteur, 'The long-term aim of socialism is not necessarily nationalisation but rather the internal transformation of the management of undertakings' (*Le Monde* 30 April 1966).

FRANCE

The French unions are weaker, numerically and financially,

than the British or German unions. Their organisation has been based on the locality rather than the factory, due to the state of the law which restricted their activity within the plant. After the riots and strikes of May 1968, the Government agreed to recognise trade union sections inside firms. This gave union representatives the right to move more freely in the factory, and to discuss wages and other matters, although boundaries are drawn defining particular activities and facilities.

The right to strike is guaranteed by the 1946 Constitution, which also protected strikers from dismissal because of their activity. There are laws discouraging political strikes, and also provision for arbitration (voluntary) and conciliation (compulsory) but there is an absence of effective sanctions and the strike rate has been relatively high.

The collective agreement in France has been developing its coverage over the years, where many of the agreements have been on a national level, and the plant agreement is now becoming more common. Yet many workers have their wages increased by legislation rather than through trade union action. In past years, the national or regional agreement has overshadowed the works agreement, which tended to be limited in scope.

Legislation about works councils (*comité d'entreprise*) was passed during the Liberation, in February 1945, and has been amended several times since. The councils must be set up in firms and enterprises with at least 50 employees. The number of representatives is based on the number of employees, with three for 50–75 employees, five for 100–500, and so on up to eleven for over 10,000. Representatives are elected by the whole staff, over 21 years of age and with at least one year's service with the firm. The term of office is for two years, with a chance of re-election. Candidates can only be presented by 'representative trade unions', where they exist; if the union lists do not have a majority, a second ballot can be held with new lists. The union lists are meant to show that the candidates are independent of the employer.

The link between the union and the works committee was altered by the 1968 Act. The delegates are now appointed by the union, and the union decides their term of office. This should mean greater influence for the union on the elected committee, as it was confined previously to an advisory role. The head of the firm acts as chairman of the committee, or appoints a deputy. The powers of the works committee include the right to be consulted on many aspects of the management of the firm, which in Britain are regarded as management matters. Information has to be given on finance and production, and an annual account is provided. Matters such as the works' rules, changes in hours of work and holiday rotas, and redundancy arrangements have to be laid before the committee. Other committees or sub-committees deal with safety, welfare and hygiene.

Facilities have to be provided for committee members, such as premises and necessary equipment. Papers and leaflets can now be distributed. Members are allowed 20 hours monthly, with additions, for their duties. They are protected from dismissal, subject to the agreement of the committee.

Profit-sharing was introduced in 1967 by the Gaullist government as a step towards 'participation'. This applies to firms with over 100 employees, on the principle that it is 'essential that employers and wage- and salary-earners, who together further the development of firms, should share the rewards of their joint efforts.'

The method of operation is that a special reserve fund is set up, and part of the profits of the firm is funnelled into the fund. This may be issued as shares, or invested as a fund inside or outside the firm. Payment to workers would normally be at the end of five years. By 1970, there were over $2\frac{1}{2}$ million workers under profit-sharing schemes out of a possible 4 million. The methods and the amounts involved are usually negotiated with the works councils.

GERMANY

The German labour movement has a history nearly as long as the British, but had a closer connection, through Karl Marx, to the mainsprings of socialist thought in the nineteenth century. There was the development of friendly societies, dealing with unemployment and sickness benefit, although this trend was hampered by Bismark's anti-socialist legislation and the begin-

186   PARTICIPATION IN INDUSTRY

nings of Germany's social security scheme.

A militant, Marxist union movement was thrown into confusion at the outbreak of the First World War, which they supported at first, though the left wing opposed the war towards the end. There were riots and dissension in the first years after 1918, then a period of comparative calm and some social reform under the Weimar Republic. A Works Councils Act was passed in 1920, stating that a council had to be set up in each establishment with at least 20 workers. The trade unions had the right to a representative, and the councils were intended to promote better understanding between employers and workers and improve productivity, as well as watching the collective agreements which had been negotiated. The objections to works councils were similar to those made forty or fifty years later; that they did not have sufficient powers, a complaint which came mostly from the left wing, while the trade unions tended to be suspicious of a development which might place power in the hands of shop stewards instead of in the national or regional union.

Although works councils survived until the early 'thirties, they were badly affected by the climate of the 1920s; the Occupation, the massive inflation, the growth of large Communist and Nazi parties, the Depression. At a time when there were talks on increasing the power of councils by giving them more information, and a law was passed in 1931, the Nazis came to power in the early 1930s, and the independent Labour movement disappeared into the Nazi Labour Front almost overnight. Numbers of active trade unionists went to prison or labour camps.

After 1945, the Allied authorities in West Germany set about rebuilding the union movement and a meeting was held in Hamburg in May 1945. Mindful of the internal disunity in the early 'thirties which had helped to split the movement, union leaders agreed to form a united labour movement, instead of splitting into religious and political groupings in line with their neighbours in France, Italy and the Low Countries.

Structure

The reconstituted unions re-grouped themselves on industrial lines, in a relatively few industrial unions. With a labour force slightly larger than that of Britain (25 m.) the Germans have one-third or eight million workers in unions. The great majority of these are in 16 industrial unions which are affiliated to the DGB, which is their equivalent of the TUC. A few unions such as civil servants and salaried employees are outside the DGB, with a million or so members. There is also a Christian Trade Union Federation, an exception to the non-sectarian approach of the DGB, which has 200,000 members. The DGB, like the TUC, is a representative body, and the unions formulate their own wage policies and negotiations.

The largest union is the Metal Workers with some two million members, covering a number of industries from steel to ship-building, and including carworkers. There are also large unions such as the Public Services and Transport Workers, the Chemical Workers and the Building Workers. The industrial unions include white-collar workers in large numbers. The important difference between British and German unions is that there is far more plant bargaining by the former. The German union is usually outside the plant, and has only recently been allowed to penetrate the work situation.

The normal collective agreement in Germany is made on an industry-wide basis. There is usually a peace clause asking for no strikes during the period of the contract, which means also no lock-out on the part of the employer.

There are organisations of employers, usually by industry, trade or region. They negotiate several thousand agreements yearly with unions. The Federal Union of German Employers' Associations is known as the BDA.

The Works Constitution Act of 15 January 1972[4]

The above Act, or *Betriebsverfassungsgesetz*, is a new addition to the lengthy history of German legislation on Works councils. The basic structure remains much as before; works councils 'shall be elected' in firms with at least five employees who have the right to vote; the number of representatives varies with the number of workers, from one for 5–20 employees to 11 for 601–1,000, and so on up to 25–35 for more than 9,000 employees and representation is now increased with the new Act; there is separate representation for wage-earners and salaried employees, and election is by all workers in permanent employment who are over 18 years. Arrangements are made for the representation of minority groups; elections are held every three years between March and May; the ballot is secret.

Those employees who can vote may present lists of candidates which have to be signed by at least one-tenth of the voting group. Disputes over elections may go to the labour court if sent by three or more workers, a relevant trade union or the employer. There are rules against sharp or questionable practices during the election, the cost of which is carried by the employer.

The new Act gives more power to works councils and places restrictions on the activities of employers, e.g. an employer may be taken to the labour court if he 'has grossly violated his duties under this Act' and the court may impose a fine for each violation up to a maximum of DM 20,000 (over £2,000) (Div. II, Sec. 3).

The works council meetings are usually held in working hours and an official from a trade union represented on the council may now be invited (1972). The separate roles of the union and the council are clearly defined, although the unions have managed to have a high representation of council delegates in some industries.

Loss of pay of delegates is made up, if they are on necessary council work. Jobs and pay levels are also protected. Provision is made for workers attending educational and training courses, subject to the 'operational requirements' of the undertaking. Delegates may have paid release on such courses up to three weeks, four in some cases for new delegates or young workers. There are limits on the numbers of members who can be given leave from duties. The works council is consulted on the above matters.

The works council receives comments and suggestions from department and works meetings. These relate to matters concerning the firm, which can include 'collective bargaining policies, social policy and financial matters'. The act states that the council shall meet at least once a month and continuous

Acts of industrial warfare between the employer and the works council shall be unlawful; the foregoing shall not apply to industrial disputes between collective bargaining parties. The employer and the works council shall refrain from activities that interfere with operations or imperil the tranquillity of the establishment. They shall refrain from any activity within the establishment in promotion of a political party (Sec. 74).

The next section carries the injunction about political activity into a stern instruction for non-discrimination. Many British employers and workers, from Smithfield meat porters to dockers, would blink at the following principles (assuming that the closed shop was ever breached).

The employer and the works council shall ensure that every person employed in the establishment is treated in accordance with the principles of law and equity and in particular that there is no discrimination against persons on account of their race, creed, nationality, origin, political or trade union activity or convictions, or sex . . . the employer and the works council shall safeguard and promote the untrammelled development of the personality of the employees of the establishment (Sec. 75).

The statements are also included under the 'General duties' of the council, which is instructed to 'promote the rehabilitation' of disabled and other handicapped workers, as well as the employment of elderly workers. To encourage youth delegates and 'the integration of foreign workers' and 'to further understanding between them and their German colleagues'.

The intention of the Act is not to reduce the rights and agreements reached by trade unions, and this is underlined in Section 77 which states 'works agreements shall not deal with remuneration and other conditions of employment that have been fixed or are normally fixed by collective agreement' unless the agreement has allowed supplementary works agreements (Sec. 77, (3)).

One of the main objections of employers to discussing financial and production matters with employees at a works council meeting, is dealt with; members of the council shall not reveal

or make use of confidential information, even after they leave
the council (Sec. 79). At the same time the employer has to
supply 'comprehensive information' to assist the council to
function properly, e.g. they may inspect the payroll showing
gross wages and salaries.

The information which the employer must give the employee
includes the nature of the job and the firm. The employee has
a right to 'be heard and request explanations'. He may ask for
the presence of a works council member, who may only make the
case public if the employee allows him to. He has the right to
see his personal file. All workers have the right to complain or
call on works council members, without prejudice. Works
councils can also hear grievances and 'induce the employer to
remedy them'. There is provision for conciliation if employer
and council fail to reach agreement on this, and other matters.
At a higher level, disputes can be referred to a labour court.

Issues co-determined by the works council

The council deals with a range of matters unless this is for-
bidden by legislation or the collective agreement; the conduct
of employees; hours of starting and finishing, as well as break
times and workdays in the week; reducing or extending the
normal hours worked; where and how payment is made;
holidays, safety and health, accommodation. There are also
matters such as pay methods and the introduction of new pay
methods, as well as work measurement.

The council has to be consulted on construction plans or
alterations, technical plant and working operations, as well as
jobs. In short, that which affects the work process and its
effects on employees. The pace and intensity of work shall be
considered. Manpower planning and future needs, vacancies,
employment and selection, grading and dismissal are discussed
and criteria or guidelines laid down. The employer and the
council shall also encourage vocational training in the establishment. Failure to provide this by the employer, as defined by
the Vocational Training Act, may lead to action by the labour
court and a maximum fine of DM 20,000. Information must be
supplied to the council on individual staff movements, temporary transfers and dismissals. The council may object to

dismissals for any or several reasons, and the worker concerned can bring an action under the Protection against Dismissal Act 'for a declaration that the employment relationship has not been dissolved by the notice of dismissal' in which event the employer has to retain him until a decision is made. The works council can uphold the dismissal and also remove employees who cause trouble in the firm.

The system of co-determination in the coal and steel industry is one of Germany's best known aspects of industrial relations. This arose from the desire of the Allies after 1945 to break up the powerful coal and steel industrial combinations, as well as providing an employee counter-balance to the employers and helping to spread the doctrines of democracy and 'social' capital. The post-war German government took up the idea and a law was passed in 1961 introducing co-determination in the coal and steel industries. The Supervisory Board (*Aufsichtsrat*) of the firm is made up of equal numbers of workers and shareholders' representatives and the employees are nominated by the works council. There is also a neutral member (making 11 in all). The union is consulted about the nominations, although, in a country where the proportion of workers is relatively low, this is not conclusive except in the industries where there is a high degree of unionisation. There is also a Labour Director on the Management Board (*Vorstand*) who is nominated by the union and whose position is protected to the extent that he cannot be removed by the Supervisory Board unless there is a vote higher than the majority of employee representatives (in large companies the total number could be ten). He is a full member of the Board and is usually a person with a trade union or Labour/political background. He usually deals with matters of social and personnel policy, but of course he can participate in other matters before the Board.

German employers, or their official spokesmen of the employer associations, have not been enthusiastic about co-determination and have expressed anxieties about its possible spread. They argue that employers risk their capital, or the funds of shareholders, and this risk should be directly related to their influence over the policy of a firm. Some see the development of co-determination as a form of 'back-door nationalisation' which the unions and the Social Democrats have formally abandoned on the political front.

The problem of role conflict for the Labour Director is found in other countries, such as Britain where the experiment has recently been tried in the steel industry. Legally he is a manager helping to run the firm, but he is also there to represent the employee interest. Issues arise where the two functions clash and role conflict appears. If he appears to side with management on issues which labour opposes, he is liable to be accused of being a 'boss's man'. If he supports labour too vigorously on the Board, the other directors may become alienated and not include him in their informal discussions which precede and often determine. An analogy with this difficulty might be seen in the dilemma of a Minister of Labour under a British Labour government. He has to represent the state interest, but this frequently conflicts with the views which the unions have of their interest, particularly in the field of wage demands in an inflationary situation, coupled with the argument about sanctions over the activities of unofficial strikers. Much seems to depend on the background and ability of the person himself; taking part in executive decisions in a large and complex industry needs either years of training, concentrated study or a deliberate decision to specialise on some narrow aspects in which the Labour Director has some experience, such as manpower planning, social and personnel policies. This is what usually happens. There is also the difficulty for some workers of adjusting to a managerial environment, but class and social attitudes have been changing rapidly and an able though inexperienced man can usually be accepted by the other directors. He remains a member of his union although he may also belong to an association for social plant management (*Verein für Soziale Betriebspraxis*).

While Labour Directors may feel they do their job successfully, and this is agreed by those in the best position to observe them, it does not follow that the men they represent have the same confidence in them. In Germany, as in other countries, there is often scepticism and distrust of union leaders and officials, which partly explains some of the unofficial strikes which occur in some countries. If this is true of union officials

who only meet management to negotiate for the men, the distrust will be stronger if the men feel that the labour director is
a part of management. One German expert writes 'There is
some danger that members of the working class advancing to
highly regarded, rather well-remunerated positions may become
alienated from their colleagues'.[5] The German unions try to
minimise the high earnings level of labour directors by asking
them to contribute to a trade union fund, but Furstenberg says
that the response has been 'only partially successful'.

Surveys of workers' attitudes to, and knowledge about, co-
determination underline the gap which exists between the
representatives and the represented and the answers tended to
reflect how the worker felt about his job rather than his views
about the system. It has been argued that the existence of co-
determination leads to better consultation and preparation for
technical change and re-training, but British experience shows
that this depends greatly on the state of the economy and alternative work, either internal or external to the firm. However, the
DGB have been campaigning for an extension of co-determination for several hundred of the larger firms and a widening in
the range of issues discussed by the Boards. The Biedenkopf
Commission, which reported favourably in January 1970,
suggested that trade unions could put candidates forward.

The basic dilemma which remains unanswered is the extent
to which the worker feels he is represented or involved in the
making of decisions which affect him. The question may have
been partly answered as the Social Democrats campaigned on
this issue, amongst others, and their election victory in late
1972 will see an extension of co-determination in the Federal
Republic, as well as a strengthening of the principle in the
EEC in Brussels.

ITALY

Unions in Italy began in the mid-nineteenth century and
regional organisation developed in the 1880s. By 1900 there
were nearly half a million union members and the General
Confederation of Labour was formed (CGL). The Catholics
developed their own workers' movement on parallel lines and,

with the approval of the Church, founded a strong national union group, the Italian Confederation of Workers (CIL), which claimed to be politically independent, whereas the CGL supported the left-wing parties.

Fascism came in the early 1920s and the corporate state eventually absorbed the free trade unions, with each industry having one fascist union.

As in Germany, the unions were restored when fighting ended in 1945. The new confederation, Confederazione Generale Italiana del Lavoro (CGIL) was intended to unite all trade unionists under one organisation. The development of the Cold War and the presence of a large Communist group in the CGIL led to its break-up and the Social Democrats and Christian Democrats left and formed their own federations. There are three main trade union groupings today.

Works councils (*commissione interne*) were revived after the Second World War and the law has been revised several times since the late 'forties. The works council constitution is similar to the German one but there must be at least forty workers in the firm compared to the German five. Trade unionists submit lists of candidates, as can groups of non-unionists. There is separate representation for manual and white-collar employees, and those over sixteen may vote, although workers must be over eighteen to be representatives, and have six months service. As in Germany and other West European countries, the number of representatives is proportional to the number of workers in the firm, with one for forty, through seven for 501–1,000, up to 21 over 40,000.

The works councils reflect the splintered state of Italian trade unionism and numbers of representatives may believe more in confrontation than co-operation. This varies from industry to industry, depending on the degree and effectiveness of unionisation. Councils without union backing are weak, and some employers are alleged to tolerate councils in order to exclude entry to trade unions.

The functions of the council are to consult with the employer about the efficiency and organisation of the firm, and to scrutinise labour contracts, along with legislation on safety and hygiene, as well as welfare. Complaints from workers are dealt with as far as possible, or they may be forwarded to the employer.

In contrast to Germany, the Italian government is rather cool towards the concept of co-management, and this view is shared by employers and to some extent by trade unions. The unions feared that strong and successful works councils would weaken their attempts to recruit and expand inside a firm. This resulted for a time in politicians opposing legislation on works councils on the instructions or wishes of the unions. Recent statutes in 1969 and 1970 recognise the existence of the councils and talk of them as 'union representation at the place of work'.[6] The statutes have gradually strengthened the position of the union in the plant, and trade union fears of docile councils in a paternalistic framework have not proved true. Further legislation in 1972 has helped to strengthen the position of the worker in the plant; but the Italian unions are still sceptical of co-determination on the German model.

BELGIUM

Trade unions grew along with industrialism, particularly from the beginning of the century, and have always been a powerful force in the country. After the First World War Joint Industrial Councils were set up, as in other countries, with equal numbers of workers and employers representatives on the committees, and chaired by a nominee of the Minister of Labour. Their functions were to deal with wages and working conditions, settle or conciliate disputes and also to provide information to the Government to improve existing legislation and to introduce new.

The JICs are the main negotiating bodies in Belgian industry and they negotiate on an industry-wide basis. They also deal with complaints or disputes which come forward from works councils. JICs (or as they are known in Belgium, *commission paritaire*) were set up in the post-1918 period, then lapsed during the Second World War, and were revived after 1945. Other bodies which symbolised the new post-war spirit of co-operation were the Central Economic Council (1948) which was intended to bring unions and employers together to advise the legislature on industrial legislation and planning. The National Labour

Council (1952) also advises ministers on social and economic matters. It adjudicates on differences between the various JICs and overseas national agreements relating to such matters as works councils. This is the structure of participation at the national level.

As in other West European countries, excepting Germany, the unions are split on ideological lines. The largest is the General Federation of Labour (FGTB) which has nearly a million members, and is socialist in the social democratic tradition. Close behind in numbers comes the other large group, the Confederation of Christian Unions (CSC). There are two smaller groupings, the staff workers' union (CGSLB) which is liberal in politics and an independent group of some half million (SCU).

The FGTB has recently reaffirmed (1971) its support for works councils and greater participation. At the same time it opposed the idea of co-determination, as have the Christian unions. This shows that, moderate though the Belgian unions appear to British left-wing eyes, they still reject the German policy of co-operation with the employers.

Works Councils date from the Acts of 1948 and 1949, and may be set up in factories with fifty or more members. They must be set up if the firm has 150 or more employees. As in other countries, there is proportional representation for manual and white-collar workers. All workers are eligible to vote in the elections for the council, although only agreed representative bodies, usually the unions, can submit lists of candidates. The numbers of representatives follow the pattern of other countries with four for 501–1,000 workers and so on up to ten for 1,001–2,000 and eighteen for 6,000. The employer is usually the chairman and he appoints representatives on the management side of the council.

The council is mainly consultative, can suggest methods of improving efficiency, and it also watches over the working of labour legislation, especially in the field of conditions of employment. The employer has to provide relevant information on the operation and finance of the company so that discussions can be fully informed. Hiring and firing can also be looked at by the council, as well as social welfare.

Safety and hygiene are usually dealt with in all but the smaller firms by a separate committee. They were set up in 1952 and are organised on the same basis as the works councils with equal numbers of representatives. They police the safety and health measures and run information campaigns.

Unions in Luxembourg are organised on the same lines as those in Belgium and they have recently been pressing for joint works councils in firms with over 250 people which would deal with promotion, dismissal and staff transfers, working and production methods. They are also asking for more detailed financial information from the employers to assist them in their negotiations.

Luxembourg takes the view that trade unions cannot represent all the people in a particular occupation; this is done by an organisation known as an occupational chamber, which is the official link between the occupation and the government and legal bodies. As Professor Als points out 'Membership of a trade union or employers' organisation is optional; adherence to an occupational chamber is compulsory.'[7] Representatives are elected to these bodies from the whole of the occupation, and their powers are consultative in relation to new legislation. There is also an Economic and Social Council which was reformed from the Council of the National Economy after the Second World War, which gave economic advice to the government. The representatives came from the occupational group 'industrialists, businessmen, craftsmen, farmers, vine-growers, salaried employees and wage earners'.[8] The Economic and Social Council was set up in July 1960 as a joint management labour committee with three government representatives. The intention of the Chamber of Deputies in setting up the Council was to extend economic democracy in Luxembourg. In a small country of 330,000 inhabitants such a body would overshadow other institutions in the field of industrial relations, and makes necessary greater participation in works councils.

NETHERLANDS

The Dutch trade union movement dates from the later nineteenth century, like that of its close neighbours. Religious divisions existed from this period, as there was an organisation of Protestant workers founded in 1877. The Catholics followed in the 1890s, encouraged by the papal encyclical of 1891, *Rerum Novarum*. One of the important trade union groups around 1900 was the National Labour Secretariat which was in the then mould of anarcho/communism. This organisation touched off numbers of unofficial strikes, although these petered out later due to lack of organisation.

The main union grouping dates from 1905, and was known as the Netherlands Federation of Trade Unions (NVV). The Christian unions also consolidated around this time and were jointly Catholic/Protestant until the Catholic hierarchy objected. They then developed separately. The groupings fluctuated in size and influence, being affected by war and the Depression. After 1945, a unified union movement foundered on the attempt by the Communists to penetrate the leadership, although the Communists are not important numerically in the Netherlands.

There are three main union groupings; the NVV, with just over half a million workers, which formed the social democrats; the Netherlands Federation of Catholic Trade Unions (NKU), about 450,000 members, which has a number of associations or clubs, apart from unions, affiliated to it, such as the Young Catholic Workers and the Catholic Workers' wives movement. There is also the National Federation of Christian Trade Unions (CNV), which is Protestant in outlook. There are unions for each branch of industry, and numbers of clubs or associations such as the Catholics have, including travel, holiday homes and legal advice. The CNV is smaller than the NKV and has 234,000 members.

The Dutch Employers have also organised, though later than the unions, in 1917. They formed the Federation of Netherlands Manufacturers' Associations, which originally set out to promote free competition and remove government regulations on industry, then began to develop an interest in social affairs. The employers divide on religious grounds as do the unions, into Protestant and Catholic, although the Roman Catholic grouping was the strongest. Curiously enough, a socialist-directed employers' association was begun in the 1930s and in 1949 was known as the Netherlands Federation of Tradespeople. The organisations are consulted on matters of social policy and the economy by the Dutch government.

The Dutch industrial scene was fairly disciplined for some years after the post-1945 period. There was widespread acceptance of wage and profit restraint which continued with varying degrees of acceptance for a number of years. This was due to the determination of Dutch employers and unions to rebuild their economy after the Second World War, and the Labour Foundation, a joint union-management body, played a leading role in encouraging co-operation and developing wage policy. The two main bodies for co-operation now are the Social and Economic Council, which sets out the criteria for wage costs and economic trends in six monthly reports, and provides the basis for discussions on the other body, the Joint Industrial Labour Council (JILC) between representatives of industry, labour and the government. Wage agreements since 1945 have had to be submitted to the Industrial Disputes Tribunal and since 1962 to the Joint Industrial Labour Council. The government also has powers to say that a collective agreement is not viable, although this seldom happens and there is ample opportunity for consultation with the parties.

This system of wage regulation has been breaking down under the inflationary pressures of the 1960s and the visible signs of higher wages and tougher union action in neighbouring countries. The government and the JILC now exercise less control over negotiations and their scrutiny is more formal than real.

Works Councils

The democratic basis of Dutch society led to early meetings of employers and workers before 1900, but works councils in their modern form stem from the post-1945 period with the 1950 Act of that name. This statute said that all firms with over 25 employees must set up a works council. Employees over 21 can vote, if they have one year's service. Workers with three years' service with the firm, who are over 23, are eligible for election. Provision is made for separate representation of manual and white-collar staff. The employer is usually chairman. Representatives increase in number with the number of employees in the firm. Elections are held every two years.

The powers of the works council are mainly consultative and are described as follows:

(a) to deal with wishes, complaints and comments brought to its attention, in so far as they affect the employees' position in the enterprise;

(b) to hold consultations regarding the fixing of holidays, work schedules, shifts, and meal breaks, if not done collectively;

(c) to ensure that the working conditions applicable to the enterprise are complied with; clauses (d), (e), and (f), state that council participants in the management of institutions attached to the enterprise, make suggestions about the firm's methods and efficiency, and supervise safety, health and hygiene measures and regulations.[9]

Lists of candidates for the councils are submitted by the unions, allowing for representation of different categories of staff. Model rules for the conduct of works councils are issued by the JILC. By the mid 1960s 45 per cent of all enterprises had a works council. This was mainly in the larger firms as 80 per cent of employees were in firms with works councils.

P. S. Pels concluded in 1966 that the various bodies for co-operation and participation, from the Social and Economic Council downwards to works councils, had created a 'climate' in which there was greater opportunity for manager and worker to speak freely. As a consequence, a more 'responsible society' had been developed.[10] In spite of this democratic framework and the good intentions of the government, situations arise in which consultation, like compassion, is not enough. Multi-national firms may decide to reduce production and close a firm down. Recent action of this kind in September 1972 led to a sit-in by hundreds of employees and a halt to the closure by the company. This action received massive trade union support and illustrates one of the 'final sanctions' left to a works council.

SCANDINAVIA

The concept of industrial democracy has been consistently supported for many years at government level in the Scandinavian countries. In 1923 a socialist prime minister, Hjalmar Branting, wrote,

> Industrial democracy thus appears as a complement to the efforts of trade unions to protect the interests of their members; but at the same time its very structure links the workers more closely to production, creating a new spur to increased productivity which is in the interests of the community as a whole.[11]

A Bill was proposed in the 1920s but was turned down by both sides of industry. The employers were against joint consultation through legislation, judging it to be an infringement of their ownership and control. The unions thought that works councils would weaken the union. In these attitudes they mirrored the responses of Western industry and labour.

By 1946 it was agreed between the Swedish employers (SAF) and the unions (LO) that works councils could be set up in firms with over 25 employees, later changed to 50 employees. The present agreement on works councils was passed in January 1967. The functions of the councils were mainly consultative, and it had been the intention of the 1946 agreement that employers would give the council (the method of selecting representatives was left to the trade unions) full information on various aspects of production, management methods and finance.

The changes in 1967 arose from dissatisfaction on the part of the works representatives, earlier examined by an LO committee which reported in 1961 on 'the trade union movement and industrial democracy'. Their theme was that the process of democratisation needed to be widened, echoing the desires of unionists in other countries for more power to the workers in shaping the decisions which affected their working lives. The employers' response was to query how far the process of joint decision-making could go without fundamentally altering the structure of management and control.

The employers, for their part, produced a working party report which was different in its orientation from that of the unions. The latter wanted more power for works councils, the former argued that existing councils did not increase the psychological satisfaction felt by the employee and that improvements might be sought which helped to develop and express the workers' personality. (See K. O. Faxen and E. Patterson, 'Labour-Management co-operation at the level of the undertaking in Sweden', *International Labour Review*, August 1967, for a fuller account.)

The final agreement made provision for wider information and discussion of labour, financial and production matters, and the greater variety of issues can be seen in a recent report by Anderman, who says that information and consultation about changes have now to be introduced at an earlier stage.[12] In Section 8 of the 1966 agreement on works councils, information has to be given to employees about planned reorganisations, installations and other changes in manufacturing, production or work methods. The LO and the SAF now have central agreements about early-warning systems. Labour market and local employment authorities and exchanges are notified for consultation and advice.

Proposals for an extension of worker representation were before the Swedish Parliament in late 1972, although employee members would still be a minority on the management board. There are also proposals for government directors in the largest Swedish companies, in order to represent the public interest.

DENMARK

Denmark has had a steady rate of economic growth, over 4 per cent in the period 1955–1964, higher than that of the UK, so that it is a prosperous country. The unemployment rate was only 1.1 in 1969, better than the British figure of over 2 per cent (which has risen since to 4 per cent in 1972). Denmark is a highly democratic country with little class or education divisions. The Conservatives and the Social Democrats are less divided over principles such as nationalisation and income distribution than is the case in Britain. This may have been due to the long period in government of the Social Democrats, who held office most of the years between 1945 and 1968.

Both the unions and the employers are well-organised. Strict control is attempted over collective bargaining as one of the 'Rules for Negotiation' of the employers' association states that 'Neither the organisation, nor their individual members, nor individual firms' can negotiate agreements on a range of issues without the general council's consent, varying from general wage increases to holidays.

Unions are organised on a craft or work basis and there is a strong shop steward system, which is protected by legislation.

Works councils were begun in the early 1920s, but their main growth, as in other countries, came after 1945. These spread rapidly after meetings and consultations and were established on a voluntary basis. The 1964 'Agreement on Joint Consultation Committees' stated that these should be set up in firms of over 50 adult employees, if demanded by the employer or a majority of the workers. The functions of the committees were to be mainly consultative, dealing with productivity and general efficiency, industrial and technical change, along with the usual themes of safety and welfare. There is a distinction made between the tasks, rights and duties of the union, and those of the committee. Agreements negotiated by the union are usually dealt with by the union.

There was an extension of the joint committee system in 1971, as had been done in Norway. Co-operation committees were formed with the intention of increasing participation and job satisfaction, and sharing responsibility to a greater degree. Further extensions of 'economic democracy' were discussed in 1972, as well as proposals for the gradual transference of ownership to a Special Wage Earners' Profit and Investment Fund.

NORWAY

Norway has combined relatively high economic growth with a reasonable distribution of the wealth produced, as we might expect from a country with a rugged tradition of work and independence based on farming and fishing. The unions are organised on industrial lines, said to derive from the Norwegians who spent some years in the USA before the 1920s, often in logging camps and influenced by the IWW idea of union organisation.

The country, like Denmark and Sweden, has had many years of Social Democratic governments, following policies of high employment, progressive taxation and comprehensive social security. Education is based on the system of state schools and there are few important social differences.

Collective bargaining, like the unions, is highly centralised, although a certain amount of wage drift has occurred in some plants. Strikes are few and the unions accepted compulsory arbitration in disputes as early as 1920. Though this lapsed in the late 1920s, it was revived again after 1945, becoming voluntary arbitration in 1953. But the unions show themselves willing to conciliate, and unofficial strikers usually pay fines imposed by the Labour Court, though the sums involved are small.

An agreement between employers (NAF) and unions (LO) after 1945 established joint production committees to promote 'effective production, good hygiene conditions and vocational training'. This has been amended and strengthened since. In 1953 there were some 1,000 committees, and it was claimed that they had made much progress in the field of social welfare. During this period, and since, the unions have done much to train members for participation on the committees.

A feature of Norwegian industrial relations is the basic agreement made at intervals of several years, setting out a framework agreed between the NAF and the LO. Part B contains the Co-operation Agreement. Under this, works councils (the new name for production committees) have to be established in concerns with at least 100 employees; with less than 100, they may be set up 'if one of the parties requests it, and the main organisation of the party agrees'.[13] Members of the council must be over 21 years, with at least two years' service. The chairman is elected 'alternately by the management and the representatives of the employees' unless another arrangement is decided by agreement.

The functions of the council are to deal with finance, and members can receive accounts of the same type as are given to stockholders at the AGM. Production plans and charges, quality and development, expansion and redundancies are discussed, along with safety and health. Certain matters may be discussed under a strict condition of secrecy if the management ask for this.

Department councils exist in smaller undertakings, and discuss similar issues to those of the works councils.

There are also co-operation committees which are informed and assisted by the Co-operative Council, set up by NAF and LO in 1966. The terms of reference state that the Council shall 'assist' and 'encourage' educational measures which will promote co-operation. It also assists the development of research into this area and publishes or disseminates the results. To assist in this work the Council has the right to ask for information but, 'may not request information on trade union secrets' (p. 26). The Co-operation Council has its own secretariat, serving a Council of six members, three from each side of industry. The post of Chairman is occupied by the heads of the employer and union groups alternate years.

The functions of the Council consist of information, training and research. Useful experience from various works councils is circulated generally. Training is carried out to further principles of co-operation. Training can be carried on in trade unions or adult education centres and in organisations.

Some research has been sponsored or supported into the workings of participation. Social scientists have given information about their research in this area, which is distributed to firms and unions. The Council itself has sent out pamphlets on the shorter working week, on successful suggestion schemes, and statistics on the establishment and activities of works councils. One-day conferences have also been organised. The follow-up to conferences and training has been organised through the highly developed pattern of workers' education and the People Correspondence Schools, with some focus on the training of shop stewards. These range from the simple to the complex. The government now finance half the expenses of courses for shop stewards lasting a week. Previously, this training was financed by the unions. The unions and the employers set up a joint fund in 1970 for trade union and management education.

One progressive development arising from the tripartite interest in participation in Norway has been the studies carried out on job design and organisation. Dr. Thorsud, who took part in some of the research, has shown that studies of employee representation on company boards in some European countries showed 'the split at the bottom of the organisational chain' which created a gap between the workers on the shop floor and those on the board. It is agreed that the board may not be the best place for representation as its activities are directed to the prosperity of the firm and workers demands at top level may run directly counter to this aim by increasing labour costs.

This division of interest between management and labour poses common problems in other countries. The researchers have sought to create groups in some selected companies which could have common goals and interests. Thorsud favours starting co-operation at lower levels, or in areas where there is less conflict: 'if democratic participation is to be a reality, then it seems inevitable that this must be started at a level where a large proportion of employees are both able and willing to participate.'[14]

The researchers noted that studies in other countries had shown that many workers wanted more responsibility, and this view was shared by others higher up the skill and responsibility ladder. Experiments were carried out in the metalworking, pulp and paper industries; jobs were redesigned along with working groups so that the work should be more varied; that jobs should follow a more meaningful pattern; that workers should have more scope for judging standards; and that 'boundary' tasks should be enlarged.[15]

One experiment in a wire-drawing department changed jobs from single operation to group work. This raised productivity and pay considerably, creating a problem for management and union, as differentials within the firm were altered considerably. The end result, after much discussion, was the union decision to shift pay patterns away from work study and individual incentives over to group work and pay.

In the paper and pulp industry, the experiment gave greater responsibility to the idea of the semi-autonomous working group, making such groups responsible for their own supervision. This resulted in the job of foreman being redesigned.

Work of this nature is a national development in a country which is anxious to promote the idea of industrial democracy.

More importantly, it delves into the fundamental aspects of organisation and decision-making in the firm and seeks to build workers' participation on solid foundations.

The extension of employee participation

Side by side with the research reported above, were moves to influence Parliament (Storting) to extend the role of employees in the decision-making process in business enterprises. Following on the various reports on industrial democracy of the early 1960s, a private members' bill was introduced before Storting in April 1967 by four MPs, one of whom (Brattelli) was the Labour Prime Minister in 1972. In January 1968, the Eckhoff Committee was set up to examine the various meanings of the phrase 'industrial democracy' and to make recommendations for its extension in industry. These were made in February 1971.

There were different views on the Committee but the Government chose to agree with those who wanted a compulsory 'Board of representatives' in companies with more than 200 employees; a third of the board to be elected 'by and from among the employees of the company'. The other groups on the Committee wanted worker directors only if the majority of workers in the firm wanted this, or opposed the idea, supporting only an extension of the existing works councils based on voluntary co-operation. The board of representatives elects in its turn the management board and its chairman. The management board takes decisions relating to investment, rationalisation, reorganisation and redundancy. Smaller firms with over 50 employees are to have a third of the management board made up of employees, or at least two workers.

These changes, announced by the Ministry of Local Government and Labour in October 1971, are meant to extend the concept of democracy in civil life into the industrial sphere. The statement ends with the hope that 'the activities of an enterprise, to a far greater extent than is the case today, will be based on the co-operation and joint responsibility of the employees' (Oslo, 22nd October, 1971).

THE PROPOSALS OF THE EEC COMMISSION

As we have seen, systems of workers' participation vary in

Western European countries, but they are usually more highly developed than in Britain. Demand for greater participation is growing along with changes in the structure of authority in management, government and unions, technological change and the rise of the multi-national company. The EEC Commission has suggested that workers be represented on supervisory boards in European companies employing over 500 people. This would be on a two-tier system, where the supervisory board appoints the executive board. The two-tier systems are found in Germany and Holland. If adopted, this worker participation would give employees a much greater influence over company closures and other changes.

The proposals follow the lines of the European Company Statute, required of those companies which have establishments in more than one member state. A number of issues affecting the employees have to be decided by co-determination, with employees having one-third of the seats on the Supervisory Board, and appoint members of the Management Board. The issue is still under consideration, but is bound to have an influence on British attitudes towards worker representation. The British Government have been asked to give their views, and the TUC have made counter-proposals approving the idea and suggesting stronger representation for workers.

The future of workers' councils

This survey of participation in Western Europe shows the widespread acceptance of the principle by management and governments. Yet the way in which workers' councils were established tells us a good deal about the structure of politics and industrial relations in the various countries; West Germany, Holland, France, Belgium and Luxembourg have all established these consultative bodies by legislative acts, whereas Italy, Denmark and Norway have set up councils through collective agreements.

While it is not easy to trace the origins of legislative acts, the French action stems from the weakness and factionalism of French unionism, a situation in which changes in industrial relations have frequently come through political decisions being reinforced by law. Germany has had works council legislation since the 1920s and the post-1945 period, and with the help of the Allies who were anxious to found a democratic union movement, saw a revival of legislation to spread participation in the coal and steel industries, the traditional basic industries of Germany's economic strength. Although Belgium has one of the strongest trade union movements in terms of membership, and has many characteristics in common with the British unions, they have had a fairly close-knit relationship between political power and union strategy. This has resulted in parliamentary legislation on works councils.

Scandinavia shows the introduction of participation by negotiation, rather than by legislation. This illustrates, in turn, the more open social structure and educational system of the Nordic countries, as well as the absence of the class, religious and political divisions which have been a feature of Western European politics.

This poses the question of the best procedure for Britain, if we wish to introduce the works council system here. We have already noted that 'Britain is the only Western European country where works councils are not mandatory either by statute or by national agreement.'[16] The British tradition is that such matters are best ordered by voluntary methods and the Donovan Report discussed the system of voluntarism, which developed over the years 1870–1970, at length.[17] The Report pointed out that the formal system of collective negotiations, based on the industry-wide agreement (about five hundred arrangements, including statutory wage-fixing bodies) was giving way to an informal system based on plant bargaining by shop stewards buttressed by full employment and tight labour control, showing their strength by unofficial and effective strikes. The Report was the first detailed inquiry into British industrial relations for fifty years, so that its comments on workers' participation must be seen as authoritative, if not decisive. This was dealt with briefly for two reasons; the first, which was the central conclusion of the Report, was that the reform of collective bargaining was the key to fruitful change and participation was of less importance; secondly, the Commission were unable to agree about participation. The TUC suggested to the Commission that workers should have representation at three levels: plant committees, regional or other levels in the enterprise, and 'legislation to allow companies, if they wish, to make provision for trade union representation on boards of directors. the TUC seek no compulsory powers, and wish to progress on a voluntary basis.'[18]

The Commission are critical of workers' participation on management bodies, and suggest instead that comprehensive agreements be negotiated which will extend and regularise the status of collective bargaining. They state that participation 'might even be harmful', arguing that managers may be misunderstood and confuse the men (this could be overcome by improving procedures or communication) and that caucuses on each side would meet before the meeting and take the real decisions (this could well happen, but it does not prevent local councils from functioning reasonably well).

The objection to worker directors is the theme of role conflict. The representative may concur in a management decision then join his mates on strike; he would be loath to support redundancy or closures; and participation 'would divert attention from the urgent task of reconstructing . . . collective bargaining'.[19] Lastly, workers' representatives would have little real influence.

As we said earlier, this view was authoritative but not decisive. The Labour Government of 1964–70, which received the Report, disagreed with it. The 1970 Conservative Government disagreed even more, rejected voluntarism, and brought in the 1971 Industrial Relations Act, the greatest legal change in the twentieth century. Voluntarism is now officially dead, and is unlikely to rise like Lazarus if a new Labour Government were elected.

Failing collective negotiations on this issue, which seems unlikely for some years, the next move should come through government legislation, on models drawn from the best of Western European experience, chief of which is the German, though the Norwegian experience is also valuable.

Once they have been convinced that the unions and the structure of collective bargaining will not be weakened, the Labour and Conservative Parties, certainly the former, might be persuaded to pass legislation requiring companies above a certain size to set up works councils. The Liberal Party, unlikely to capture power, often reflects future change, as they did on the Common Market issue. Their proposals include the reform of company law for public companies, making employees members with the same rights as shareholders. The board of directors would be elected, as would works councils.

Developments in Britain may also be influenced by membership of the Common Market. The European Commission, which forms policy for the Community, was suggesting in October 1972 that the member states, now joined by Britain, Ireland and Denmark, should adopt the systems of workers' representation in Holland and West Germany. These systems are two-tier, as companies have a supervising council and a board of management. The council would 'supervise the board of management on certain important issues' such as closures, changes in company employment or production policy, and other important changes.[20]

While these proposals have to be approved by the member states, they represent a far greater change for Britain than they do for the other countries, some of whom are well on the road to worker participation. There is no reason why the proposals should be unacceptable to British management and unions. Britain has now reached the point where there is need for a 'Copernican' shift in industrial relations. It may be that the change can only come through a radical political shift, but the experience of the last twenty-five years suggests that a number of agreed or compromise measures, such as the need to fight inflation, to survive and prosper in the Common Market, to share wealth and power equitably and reach out to other countries, can only be done by stages. One of these stages could be a move forward to a wide ranging statutory scheme of workers' participation.

REFERENCES

1. A. Sturmthal, *Workers' Councils*, Harvard University Press, Harvard, 1962.
2. XXI National Trade Union Congress, Hungary, May 1967, cf. *Participation of Workers in Decisions within Undertakings*,

ILO, Geneva, 1969.

3. L. G. de Bellecombe, op. cit.
4. ILO, Federal Republic of Germany 1, Works Constitution Act, from *Bundesgesetzblatt*, 18 January 1972; F. Bleistein, *Die Neue Betriebsverfassung*, Stollfuss, Bonn, 1972.
5. F. Furstenberg, *Bulletin No. 6*, International Institute for Labour Studies, June 1969, Geneva, p. 128.
6. OECD, *Recent Trends in Collective Bargaining*, International Management Seminar, Paris, 1972, p. 53.
7. G. Als, Collaboration between the Authorities and Occupational Organisations in Luxembourg, International Labour Review, Geneva, June 1966, p. 621.
8. Als, op. cit., p. 632.
9. Ministry of Foreign Affairs, *Digest of the Kingdom of the Netherlands, Social Aspects*, The Hague, 1968, pp. 47–48.
10. P. S. Pels, Organised Industry and Planning in the Netherlands, I.L.R., September 1966.
11. E. Rhenman, *Industrial Democracy and Industrial Management*, Tavistock, London, 1968, p. 75.
12. S. Anderman, *Trade Unions and Technological Change*, Allen and Unwin, London, 1967.
13. Basic Agreement of 1969 (Hove de vatelen), Oslo, 1969, pp. 20–1.
14. F. Thorsud, 'Social Technical Approach to Job Design and Organisational Development', *International Management Review*, Wiesbaden, 1968.
15. Op. cit.
16. Institute of Personnel Management, *Workers' Participation in Western Europe*, Information Report No. 10, September 1971, London.
17. The Donovan Report, p. 257.
18. Op. cit., p. 5.
19. Op. cit., p. 258.
20. *The Times*, 6 October 1972.

Bibliography

C. Argyris, *Integrating the Individual and the Organization*, Wiley, New York, 1964.

F. H. Blum, *Towards a Democratic Work Process*, Harper, New York, 1953.

W. M. Blumenthal, *Co-determination in the German Steel Industry*, Princeton University Press, Princeton, 1956.

J. A. C. Brown, *The Social Psychology of Industry*, Penguin Books, Harmondsworth, 1954.

L. Coch and J. R. P. French, 'Overcoming Resistance to Change', *Human Relations* (1), 1948, pp. 512–32.

K. Davies, 'The Case for Participative Management,' in Huneryager and Heckman, q.v.

A. Etzioni, *A Comparative Analysis of Complex Organizations*, Free Press, Glencoe, 1961.

A. Flanders, *Collective Bargaining*, Penguin Books, Harmondsworth, 1969.

Ted Fletcher, 'The Road to Joint Control,' *Management Today*, April 1970, pp. 90–3 and 162.

F. K. Foulkes, *Creating More Meaningful Work*, American Management Association, New York, 1969. Bailey Bros. & Swinfen, 1969.

Alan Fox, 'Industrial Sociology and Industrial Relations,' Research Paper, No. 3: Royal Commission on Trade Unions and Employers Associations, HMSO, 1966.

Alan Fox, 'Managerial Ideology and Labour Relations,' *B.J.R.*, November 1966.

J. de Givry, 'Developments in Labour-Management Relations in the Undertaking,' *I.L.R.*, January 1970.

J. H. Goldthorpe *et al.*, *The Affluent Worker: Industrial Attitudes and Behaviour*, Cambridge University Press, Cambridge, 1968.

F. de P. Hanika, *New Thinking in Management*, Hutchinson, London, 1965.

F. Herzberg, *Work and the Nature of Man*, Staples Press, London, 1968.

213

F. Herzberg, B. Mausner and Barbara B. Synderman, *The Motivation to Work*, Wiley, New York, 1959.

Institute of Industrial Relations, University of California (Berkeley), 'Symposium on Workers' Participation in Management,' *Industrial Relations*, 9 (2), February 1970.

J. Kolaja, *Workers' Councils: The Yugoslav Experience*, Tavistock, London, 1965.

K. H. Lawson, 'Universities and Workers' Education in Britain,' *I.L.R.*, January 1970.

Yair Levy, 'The Contribution of Co-operation and Trade Unionism to Improved Urban-Rural Relations,' *I.L.R.*, June 1966.

R. Lickert, *New Patterns in Management*, McGraw Hill, New York, 1961.

T. Lumpton, *On the Shop Floor*, Pergamon, Oxford, 1963.

Arthur Marsh, Joint Consultation Revived? Trends in Personnel Management No. 2, *New Society*, 16 June 1966, pp. 14–15.

A. H. Maslow, *Motivation and Personality*, Harper Bros., New York, 1954.

D. McGregor, *The Human Side of Enterprise*, McGraw Hill, New York, 1960.

National Board for Prices and Incomes, *Productivity Agreements*, Report No. 36, Cmnd. 3311, HMSO, 1968.

F. E. Oldfield, *The Making of Managers*, Mason Reed, 1967.

D. Pym, *Industrial Society: Social Sciences in Management*, Penguin, Harmondsworth 1968.

Eric Rhenman, *Industrial Democracy and Industrial Management*, Tavistock, London, 1968.

V. Rus, Influence Structure on Yugoslav Enterprise, Institute of Industrial Relations, University of California (Berkeley), *Industrial Relations*, 9 (2), February 1970, pp. 148–60.

J. Y. Tabb and A. Goldfarb, *Workers' Participation in Management*, Pergamon, Oxford, 1970.

V. H. Vroom, *Work and Motivation*, Wiley, New York, 1965.

Kenneth F. Walker and L. G. de Bellecombe, 'Workers' Participation in Management: The Concept and its Implementation,' *Bulletin*, International Institute of Labour Studies, Geneva, 2 February, 1967, pp. 67–100. See also Bulletins 3–9.

Index

Administrative decisions, 139–43, 170, 178
Allis-Chalmers, Company, 173
Als, G., 197
American Management Association, 164–5
Anderman, S., 202
Antoni, A., 47–8
Associated British Tissues, 153
AUEW, 153

Barnard, C., 164
Belgium, 195–6
Bell, D., 52
Bellecombe, L. G. de, 183
Bevin, E., 144
Blum, F. H., 166
Blumberg, P., 32
British Labour Party, 9–10, 53, 160
  Levy, A. B., 164
Briant Printing Company, 173–6
British Leyland Company, 173
BP Plastics Dept., 153
British Steel Corporation, 83–107, 140
Brown, W., 141, 146, 147 148, 149
Burnham, J., 167

Campbell, J. R., 118
CBI, 90
Coal Industry, 15, 56–79
  Nationalisation Act, 56, 70
Co-determination, 70, 140, 178, 190–3
Child, J., 74
Chamberlain, N. W., 138
CIR, 11
Christenson, T., 27
Co-operative Productive Fedn., 40
Cole, G. D. H., 8–9, 28, 34, 53, 108, 179
Co-operative Societies, 138, 140
Co-partnership, 42–3, 157, 158–60, 178
Collective bargaining, 142, 145, 149–55
Communication, 57
Communist attitudes, 183
Conflict, 60
Copeman, G., 164

Cunningham, Sir G., 74

Dahrendorf, R., 15, 64
Daniels, W. W., 2, 124
Denmark, 202–3
Definitive policy, 146–8
Discipline, 59, 150–2, 153–4, 163
Directors, boards of, 139, 143, 159, 160
Directive functions, 138, 167
Dismissal, 59
Donovan Report, 4, 90, 209–10

EEC Commission, 207
  proposals for participation, 207–8
Entrepreneur, 138
Executive decisions, 139–43

Factories for Peace, 166
Farmer and Sons, 162, 166
Faxen, K. O., 202
Filson, A. W., 34
Fisher-Bendix, 173
Firestone, 154
Flanders, A., 177
Ford, Ayrton & Co., 159–60
Fleck Report, 65
Fox, A., 76, 89–90, 146
France, 183–5
French, J., 139
Furstenberg, F., 193

Gallacher, W., 118
Germany, 185–93
  trade unions, 186–7
  Works Constitution Act, 187–92
  co-determination, 190–3
Glacier Metal, 143, 146–9, 166
Glyn, A., 155
Goldstein, J., 3
Goldthorpe, J., 74
Goyder, G., 166
GKN-Shotton, 150–2, 154
Government, 64
Greening, E. O., 25–6
Griffin, A. R., 78

215

Hall, F., 42
Harvey, 33
Holdsworth, Sir H., 65
Hawley, C., 155
Horne, R., 176
Horner, A., 142
Hughes, C. L., 142
Hurst, F. A., 165

Incomes policy, 80
Industrial Relations Act, 9, 171
Industrial relations, 2
Industrial tribunals, 9
ILO, 182, 183, 187–90
IPM, 209
Italy, 193–5
    Unions and works councils, 194–5

Jaques, E., 142, 143, 146, 166
Johnson & Johnson, 105–6
John Lewis Partnership, 144, 157–8
Joint Consultative Committees, 64, 65, 66, 72

Kahn-Freund, O., 169
Kalamazoo, 160
Kaplan, H. A., 67

Lasuen, J. R., 54
Levels of decision, 139–43
Luxembourg, 197

Madden, G. H., 50–1
Mann, A., 31
Marshall, A., 37–8
Maudling, R., 181
McGarry, E., 174
McIntosh, N., 2
McCarthy, W. E. J., 10
Michael Jones Community Ltd., 159
Ministry of Labour, 157

Nat. Board for Prices and Incomes, 152
Nat. Coal Board, 56, 58, 64, 65–75, 77, 81
Nat. Industrial Relations Court, 176
Nat. Union of Mineworkers, 57
Netherlands, 197–200
    unions and employers, 198–9
    works councils, 199–200
Norway, 203–7
    joint production committees, 204

co-operation agreement, 209
co-operative council, 205
education and research, 205–7
developments in participation, 207
Northcott, C. H., 74

Owen, R., 162
Ostergaard, G., 44–5
OECD, 195

Participation, 1–2
    critics of, 12
    Common Market, 12
    Labour Party, 10
    marxist view, 4–5
    and management, 5
    political parties, 9–10
    private sector, 18–19
    single channel, 12
    and trade unions, 1–3, 19
    U.S. unions, 12
    universities, 3–4
    Yugoslav, 18
Pafford, J. H. P., 159
Parkinson, N., 141
Patterson, E., 202
Patterson, T. T., 142
Paynter, W., 71
Pels, P. S., 200
Plant closures, 146–60
Penrikyber Colliery, 72
Plessey, 173
Power loading, 64
Private enterprise, 138 et seq.
Productivity bargaining, 76, 152–4, 172

Raven, C. E., 53
Ransom, Hoffman and Pollard Coy., 173
Redundancy Payments Act, 171, 177
Rees, W. D., 69
Reid, J., 118–19
Rhenman, E., 201
Richbell, S., 153
Robertson, D., 33
Roberts, R. D. V., 74
Robens, Lord, 75
Rosenstein, E., 20

Sawtell, R., 52
Scandinavia, 201–7
    industrial democracy, 201

Shareholders, 140, 141, 154–66
Shipbuilding, 17–18
  Advisory Group, 116
  Central Worker Council (Yugo-
    slavia), 125
  development of participation, 113–
    16
  Fairfields, 108, 111–16
  Communist Party, 117–18
  Departmental Workers' Councils, 129
  industrial characteristics 108–09,
  production cycle, 110
  industrial democracy in a Yugoslav
    shipyard, 125
  management and finance, 130–2
  Upper Clyde Report, 135
    Reid, J., 118–19
    Liquidator (R. C. Smith), 119,
      120, 121
    'Work-in', 121–3
    right to work, 123–4
  Shop Stewards Co-ordinating Ctte.,
    116, 119
Strauss, G., 20
'Sit-in', 172–7
Smelser, N. J., 26
Staff status, 153
Steel industry
  job descriptions, 104–5
  management theorists, 90–1
  product divisions, 86
  worker (employee) directors, 16–17
    definition of, 84, 89
    evaluation and changes, 87, 97–100
    Group Boards, 85
    unions, 85
    introduction of, 91–3
    nominations, 88
    conclusions, 102–3
Syndicalism, 139, 141
Scott-Bader Commonwealth, 162–3, 166

Stedeford, I. A. R., 165–6
Scottish TUC, 176

Thorsud, F., 205, 206
Trade unions
  as 'permanent opposition', 7
  decentralisation, 4
  education, 11
  private industry, 167–77
  productivity bargaining, 8
Treaty of Rome, 181
TUC, 79, 83, 90

United Nations (UNIDO), 40

Watkins, W., 42
Western Europe, 182
Workers co-operatives
  contrast with co-operative co-
    partnership and consumers
    co-operatives, 42–3
  current approaches, 40
  criticism of, 32
  declining and sweated trades, 23–4
  definition of, 21
  ideological influences, 25–30
  France, 46–8
  Italy, 47
  present numbers, 45
  prospects, 50–2
  reasons for failure, 34–40
  sales, 45–6
  raising capital, 40–1
  managerial methods, 31
  expert advice, 40
  unemployment, 24–5
  historical origins, 21–32
Webb, Beatrice, 33, 37, 38–40
Works councils, 3, 154, 208–11 (see
  under countries)
Work-in, 172, 173, 176